Marijuana Grower's Guide

"The industrious pair, Mel Frank and Ed Rosenthal, are the world's foremost authorities on pot agronomy."

—*High Times Magazine*

"Not only have these authors provided us with all information necessary to grow our own, but here too is the history of one of the most socially and botanically fascinating plants man has known . . . the disciplined nature of the writing will help put drug literature on the road to respectability."

—*Los Angeles Times*

"It is, on balance, an extremely clear and interesting essay in practical horticulture, as accessible a study of a single plant, at this high level of seriousness, as one is likely to find."

—*New York Times Book Review*

"This is a very wise and masterful book about marijuana as a plant which needs attention. If it wasn't marijuana, the book would find a welcome place on a flower-fern grower's shelf and most certainly will appeal to professional and amateur botanists."

—*West Coast Review of Books*

Marijuana
Grower's Guide

Deluxe Edition

by Mel Frank
and Ed Rosenthal

And/Or Press
Berkeley • California

Micrographs taken with an optical microscope. Illumination is by incident sunlight or daylight. Most exposures are with Kodachrome 64, 35mm film. Magnification is given to the negative. Photographs and micrographs by Mel Frank.

Cover photograph: Hi Leitz
Design: Bonnie Smetts
Typesetting: Media Matrix
Paste-up: Paulette Traverso

© 1978 by Mel Frank and Ed Rosenthal

Published and Distributed by
And/Or Press
P.O. Box 2246
Berkeley, CA 94702

ISBN: 0-915904-26-8
Library of Congress Catalog Card No.: 77-82452

Printed in U.S.A.
First printing—September, 1977
Second printing—April, 1978
Third printing—November, 1978
Fourth printing—April, 1979
Fifth printing—July, 1979
Sixth printing— April 1980

The material in this book is presented as information which should be available to the public. The publisher does not advocate breaking the law. However, we urge readers to support N.O.R.M.L. in its efforts to secure passage of fair marijuana legislation.

"It is a plant—this thing that we are about to discuss: a green plant, a very abundant and ubiquitous plant, an unusually valuable economic plant, possibly a dangerous plant, certainly in many ways a mysterious plant."

Dr. Richard Evan Schultes,
Random Thoughts and Queries on the Botany of Cannabis[14]

The plant is Cannabis: *the weed we know as the marijuana or hemp plant.*

"Weeds are not restricted to the plant kingdom. Numerous animal species have weed races. The housefly is weedy across the world, rabbits are weedy in Australia, and the holy Brahman cow is weedy in Hindu India. Indeed, man is the ultimate weed, being obligately confined to the habitat he is creating."

J. M. J. DeWet and J. R. Harlan,
Weeds and Domesticates:
Evolution in the Man-Made Habitat[207]

Contents

Foreword

Throughout recorded history, marijuana has been with us. But only in the past fifty years has all the fuss been kicked up about it. Marijuana has been used as food, fiber, and medicine. Monks in the Middle Ages used the seeds in porridge, early American settlers wove it into fabric, and for centuries folk medicine extolled its virtues.

That has all stopped. Diets, even for monks, have changed. Science has fashioned new synthetic fabrics. And doctors are forbidden to prescribe medicine. Yet we are entering a new era of understanding about this ancient plant.

It is no longer a crime to possess a small amount of marijuana in ten states, and in one, Alaska, you can even grow it in private without penalty. And through research, medical science is beginning to discover the old uses of pot as well as some new ones.

In 1970, former President Richard Nixon and former Attorney General John Mitchell convinced Congress to pass the Federal Controlled Substances Act, a complete overhaul of the nation's drug laws. Part of their scheme classified marijuana in "Schedule I" along with heroin, and succeeded in making a political statement rather than a judgment based on scientific knowledge. They were interested in reinforcing older Americans' fears and misinformation concerning marijuana's potential to harm the user, and in particular its purported ability to "lead to harder drugs." The result of this politicalization of science is that the prohibitions that apply to heroin and its use also apply to marijuana. One of those limitations is that heroin (and now marijuana) is deemed to have no medical usefulness, and thus is not available for a physician to prescribe regardless of demonstrated need. A physician who attempted to prescribe marijuana to a patient for what he believes to be a legitimate medical application would be subject to the same criminal penalties as would someone who sold a quantity of marijuana on the illegal market. He would be considered a "pusher" and would be chargeable with a five-year felony.

But as part of the massive research effort to try to find something wrong with marijuana, scientists and the Federal government stumbled across the fact that it has several important medical applications. Though the government still has not documented any significant harm to the user, this rediscovery of marijuana's potential therapeutic value has now led government to increase research into this positive side, and the findings continue to mount.

NORML first filed a suit against the government in 1972, attempting to force the reclassification of marijuana. We knew at that time the government would probably not totally remove marijuana from Federal control, although that was one of our requests. Nonetheless, the process of re-examining marijuana was a process we knew the government could not endure without eventually rescheduling marijuana to the extent that it would be available by prescription when appropriate. Following five years of legal maneuvers, including two appeals by NORML to the U.S. Court of Appeals, the Drug Enforcement Administration (DEA) has been forced to refer our request to the Department of Health, Education, and Welfare for the purpose of holding hearings on marijuana's medical usefulness. With the evidence already in, we expect that marijuana will soon be classified to permit its important medical applications once again.

Although a legal reclassification of marijuana will correct one inequity, it will not alter the fact that several hundred thousand otherwise law-abiding people in this country are needlessly and often tragically arrested each year. These are people, like myself, who use marijuana recreationally. According to the government's latest surveys, 35 million Americans have now tried marijuana, and 15 million are regular users.

Though marijuana smoking only gained the attention of mid-America in the late 1960's and early 1970's, the reader will discover marijuana's long and varied history in the first two chapters of Mel Frank and Ed Rosenthal's updated work. Only since the 1920's has marijuana been considered illegal.

RECENT MARIJUANA PENALTIES

From the 1920's through the 1960's, marijuana users were generally penalized heavily. The Federal penalties for marijuana, first adopted in 1937, were increased in 1951, under a series of amendments called the Boggs Act. It was at this time that the concept of minimum mandatory sentencing was first introduced under Federal drug legislation, something which Nelson Rockefeller later tried when he was Governor of New York as a "new approach." It failed to stem the use or sale of drugs in either case.

In 1956, the Federal government raised penalties even further, spurred on by reports of increased marijuana smoking. It is important to note that the response of the Congress and the state legislatures to this point was uniform: penalties were increased periodically, as marijuana smoking became more prevalent. It was after the 1956 amendments, during the time when the United States had the strongest penalties ever, that marijuana smoking went through its most significant increase. Despite harsh felony penalties in almost all jurisdictions, marijuana smoking became significantly more popular in the 1960's. The criminal law demonstrated irrefutably its ineffectiveness as a deterrent to drug usage.

MARIJUANA AS A SYMBOL

Associated with the anti-war movement in the late 1960's, marijuana smoking became a symbol itself, a rejection of the traditional culture's use of alcohol and a symbolic gesture in favor of new values. Its use became prevalent for the first time among large numbers of middle-class college students and others active in social protest.

As a result, marijuana became a negative symbol to more traditional Americans, feared by the establishment more as a perceived threat to their value system than as a health threat. It was somehow feared that marijuana smoking caused people to become radical in their political thinking, that it would lead to the demise of the established value system in this country. The response was "get tough" and the number of marijuana arrests rose significantly. From a low of 18,000 in 1965, marijuana arrests rose tenfold by 1970, and 441,000 occurred in 1976. No longer are the vast majority of those being arrested minority-group members, but white and middle-class. Not surprisingly, the legislatures began to respond for the first time to moderate the penalties rather than to further increase them. When it was someone else's kids who were smoking marijuana, they were criminals who should be punished. But when it became our kids, and we had to decide whether our kids were criminals or whether the law was wrong, it was easier to conclude that, in this instance, the law is indeed an ass.

1970 FEDERAL CONTROLLED SUBSTANCES ACT

So in 1970, as part of the Federal Controlled Substances Act passed by the Nixon-Mitchell team, Federal possession penalties for all minor drug-possession offenses were lowered from a felony to a misdemeanor. That, of course, is not decriminalization, since an individual still remains subject to being arrested and jailed for up to one year and/or fined $5,000. Even though this was the same law that misclassified marijuana with heroin, it did represent a change in direction.

STATE REFORMS

This limited Federal change in attitude prompted states to begin to re-examine their harsh felony penalties as well. To date, only Nevada retains felony penalties for minor marijuana offenses, and Arizona retains a law which permits the prosecutor the option of proceeding with a case as either a felony or a misdemeanor. All other states have lowered their laws at least to a misdemeanor level (usually defined as up to one year in jail). As of September 1977, ten states have stopped arresting marijuana smokers. In these states (Oregon, Alaska, Maine, Colorado, California, Ohio, Minnesota, Mississippi, New York, and North Carolina) the legislatures have generally followed the recommendations of the National Commission on Marijuana and Drug Abuse (Schafer Commission). In their comprehensive report, "Marijuana: A Signal of Misunderstanding," released in 1972, the Commission surprised almost everyone in the country by failing to support then-President Nixon's preconceived notions that marijuana was a killer drug. The Commission held hearings all around the country, and heard from medical experts as well as marijuana smokers. The 13-member Commission concluded that, although marijuana smoking should be discouraged as a matter of national policy, as should all recreational drug use, its use did not present any significant threat to the user or to society. They recommended that possessing small amounts of marijuana should not continue to be defined as a criminal matter.

THE OREGON MODEL

Of the ten states which have thus far adopted decriminalization measures, nine have opted for a modification first used in Oregon. Since the law became effective in October, 1973, marijuana smokers have been subject to a maximum $100 civil fine. Enforcement is with a traffic-ticket-like citation instead of an arrest. An individual is not taken into custody, does not get a criminal record, and is not threatened with any jail term. The person is treated in essentially the same way as a person caught for a driving violation.

This model plan has been so successful that eight other states have adopted it, and many others are currently considering similar changes.

THE ALASKAN MODEL

Though the Alaskan state legislature approved a system similar to the Oregon plan, retaining a maximum $100 fine even for private marijuana smoking, the Alaskan State Supreme Court in May of 1975 overruled part of that new law. In a unanimous decision, the Court held that the state constitutional guarantee of right to privacy precludes any penalty for private marijuana possession or use by adults. *Ravin* v. *State*, 537 P.2d 494 (1975).

As a result of the *Ravin* decision, there is no longer any state penalty in Alaska for possession of marijuana in private. The Court's decision is especially significant because it recognized that the old myths about marijuana have not stood up under scrutiny. The Court pointed out that current medical and scientific reports on marijuana have shown it to be a drug which poses no substantial risk to the public health, safety, or welfare. The Court recognized that the fundamental rights involving the privacy of the home, and of individual autonomy, outweigh the government's interest in prohibiting this relatively innocuous drug.

NORML has been involved in a number of important constitutional and other issues in the courts, and we now have a number of cases pending in state and Federal courts seeking decisions similar to the one in Alaska. NORML is bringing constitutional challenges, based on the right to privacy, in New York, California, Illinois, the District of Columbia, and a number of other states. We fully expect that other states will eventually follow the far-sighted decision of the Alaska Supreme Court.

NORML supports the Alaskan model, which provides for *no* penalty for private use.

CULTIVATION

As a result of the privacy ruling, Alaska is currently the only state which permits private cultivation of marijuana for personal use. Though the climate in Alaska is hardly optimum for growing marijuana, the consumer in that state at least has the legal option of growing his own, a concept which will undoubtedly spread.

In the other states where possession of a small amount of marijuana has been decriminalized, efforts are now underway to decriminalize private cultivation of marijuana for personal use. Presently the states of California and Oregon are considering legislative proposals, and several other legislatures will soon be doing the same. NORML is in total support of this concept, as part of the individual's constitutionally protected right to privacy, and as a practical alternative to the black market for those who wish to grow marijuana for their personal use. We recognize that many people prefer convenience, and will not grow their own marijuana. But at the same time, many people object to dealing with the underground market, either in principle or because of the possibility of being exposed to adulteration, "ripoffs," and other, potentially more dangerous drugs. For those individuals, growing a green plant in one's closet or back yard should be permitted. Certainly the potential for abuse is negligible in that situation.

Finally, cultivation should be recognized as an important element in the growing drive to effectively organize the marijuana consumer. NORML has long attempted to represent the perspective of the consumer—the marijuana smoker—in this issue. It is not our wish or policy to encourage use, but rather to speak in a responsible manner for the marijuana smoker. We believe that one of the dangers which lies ahead is the possible exploitation of the marijuana market by major commercial interests, including the giant cigarette companies. One of the most effective levers—perhaps the only effective one—that the consumer has against that eventual takeover is to guarantee that the individual is always permitted the option of growing his own marijuana. Thus, if the major companies offer marijuana which is too weak, which is too highly taxed, which is so commercialized by Madison Avenue as to offend us, we can simply refuse to buy their marijuana and grow our own instead. And although those considerations may now seem futuristic, within the next few years they will become real and important.

Keith Stroup, Esq.,
National Director,
National Organization for the Reform
of Marijuana Laws (NORML).

Washington, D.C.
Fall 1977

Preface

The purpose of this book is to show you how to grow enough marijuana to supply all your family's needs. It doesn't matter where you live, or even if you are growing your first plant, because all the information needed to become a master marijuana farmer in your own home, or in the field, is provided in these pages.

The world has seen an enormous increase in marijuana use in the past ten years. Consequently, many governments have sponsored research in order to understand the nature of the plant as well as its psychoactive compounds—substances that are being smoked or ingested by more than 400 million people all over the world. Before this recent interest, marijuanaphiles had only a few research papers (mostly on hemp varieties) to glean for information about the plants and their cultivation. Now there are thousands of papers dealing directly with the plants and their use as marijuana. This doesn't mean all is known about marijuana. In fact, much of what is discussed deals with unknown aspects of these ancient and mysterious plants. The mysteries, however, are beginning to unravel.

Our information resources include our personal experience with growing and the experience and knowledge shared with us by marijuana growers all across the country. We also rely on the professional research of many scientists (see the Bibliographic Notes). For the experienced growers, we've included the latest research on increasing potency, some ideas for improving yield and controlling flowering (time of harvest), and also procedures for breeding quality strains suited to a particular growing situation.

Some of the best grass in the world is grown right here in the United States (that is our very own stoned opinion of homegrown gratefully sampled from Hawaii to Maine). You can do it too—it's not magic, and it's not difficult to do. Highly potent plants can be grown indoors, as well as in gardens, fields, and the wilds. Indoor growers must create an environment, whereas outdoor gardeners work within the environment. Following these two approaches to cultivation, this book is divided into separate, parallel parts on indoor and outdoor sections, preceded by some background information on marijuana plants, and followed by general procedures for breeding, harvesting, etc., that are independent of the type of growing site.

Cultivation is not a complicated process, and we hope we don't make it appear difficult. But even if you're a novice when you first sow your seeds, your questions on the plants and their cultivation will become more complex as you gain experience and insight. We hope we have anticipated your questions with solid and clearly stated answers; we intend this book to serve as a guide long after your first reading and harvest.

There are probably as many ways to grow marijuana as there are marijuana farmers. We hope to impart an understanding of the plants and their cultivation, so that you can adapt the knowledge to fit *your* particular situation—where you live, the land or space available, and the time, energy, and funds at your disposal.

Modest indoor gardens are quite simple to set up and care for. All the materials you'll need are available at nurseries, garden shops, and hardware and lighting stores, or they may be found around the house or streets. The cost will depend on how large and elaborate you make the garden and on whether you buy or scavenge your materials. With a little ingenuity, the cost can be negligible.

It takes about an hour every three or four days to water and tend a medium-sized indoor garden.

Outdoors, a small patch in your summer garden can supply all your smoking needs with little or no expense. Generally, marijuana requires less care than most other crops, because of its natural tenacity and ability to compete with indigenous weeds. Hardy *Cannabis* resists mild frost, extreme heat, deluge, and drought. In this country, few diseases attack marijuana; once the plants are growing, they develop their own natural protection against most insects.

In some areas of the country, such as parts of the Midwest and East, the plants may require no more attention than sowing the seeds in spring and harvesting the plants in autumn. But if you're like most growers, you'll find yourself spending more and more time in your garden, watching the tiny sprouts emerge, then following their development into large, lush, and finally resinous, flowering plants.

Nurturing and watching these beautiful plants as they respond can be a humanizing experience. Marijuana farmers know their plants as vital living organisms. If you already are a plant grower, you may understand. If not, read through this book, imagining the various decisions you, as grower, would be making to help your plants reach a full and potent maturity. Then make your plans and get started. There's just no reason to pay $50 an ounce for superior smoke when it grows for free. Free grass, free yourself.

This book is the result of the efforts of many people, each of whom contributed uniquely to its final form and content. First there are the many growers who opened their hearts and gardens to us. Our love and thanks to our friends in California (Calistoga, Calavaras, Humbolt, Orange counties, and the Bay Area), the Umpqua Valley, Oregon, Eastern Colorado, Central Florida, Eastern Massachusetts, Upstate New York, New York City, Atlanta, Hawaii, and Port Antonio, Jamaica. We would also like to thank everyone who wrote and shared their growing experiences with us.

Specifically, we would like to acknowledge the contributions of the following: *Editors;* Aidan Kelly, Peter Beren, Ron Lichty, and Sayre Van Young. *Preparation of the manuscript;* Carlene Schnabel, Ron Lichty, Aidan Kelly, Marina La Palma. Index by Sayre Van Young. *Layout and Design;* Bonnie Smetts. *Graphics;* maps and charts by E.N. Laincz; illustrations by Oliver Williams; and molecules by Marlyn Amann. *Special Services;* Gordon Brownell, Al Karger, Michael Starks, Peter Webster, and special thanks to Sandy Weinstein for help with the photography. Also thank you M.T., A.P., and C.T. *Special thanks;* Sebastian Orfali and John Orfali.

We were fortunate to have had the use of the following libraries: Bronx Botanical Gardens, City College of New York, Fitz Hugh Ludlow Memorial Library, Harvard Botanical Museum, New York City Public Libraries, University of California, Berkeley and San Francisco, University of Mississippi, Oxford.

PART 1.
General Information about *Cannabis*

Chapter One
History and Taxonomy of *Cannabis*

CANNABIS AND ANCIENT HISTORY

The ancestors of *Cannabis* originated in Asia, possibly on the more gentle slopes of the Himalayas or the Altai Mountains to the north. The exact origin, obscured by Stone Age trails that cross the continent, is not known.

We don't know when *Cannabis* and humanity first met. Given the growth habit of the plant and the curiosity of humanity, such a meeting was inevitable. In the plant world, *Cannabis* is a colonizer. It establishes new territory when running water or seed-eating animals carry seed to cleared and fertile soil open to the sun. Fertile soil, clear of competing plants, is rare and short-lived in nature, and is commonly caused by catastrophe such as flood or earthslide. Natural dissemination is slow and the plants tend to grow in thick stands by dropping seed about the spread of their branches.

During the Neolithic era, some 10,000 years ago, nomadic groups scavenged, hunted, fished, and gathered plants in an unending search for food. The search ended when they learned to plant the native grains (grasses) and developed agriculture. Agriculture requires a commitment to the land and grants a steady food supply which enables people to form permanent settlements. *Cannabis* and Neolithic bands probably came in contact often as the plants invaded the fertile clearings—the campsites, roadsides, fields and garbage heaps—that occur wherever people live.

In 1926 the Russian botanist Vavilov summarized the observations of his comrade, Sinkaia, on the domestication of hemp by peasants of the Altai Mountains: "1. wild hemp; 2. spreading of hemp from wild centres of distribution into populated areas (formation of weedy hemp); 3. utilization of weedy hemp by the population; 4. cultivation of hemp."[24]

The plants which people learn to use help define aspects of

their way of life, including perceptions of the world, health, and the directions their technologies and economies flow. The plants you are about to grow are descended from one of the ancient plants that made the transition to civilization possible.

The earliest cultural evidence of *Cannabis* comes from the oldest known Neolithic culture in China, the Yang-shao, which appeared along the Yellow River valley about 6,500 years ago.*

Figure 1. Hemp stems.

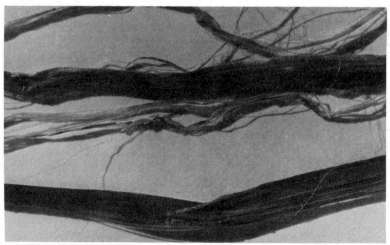

Figure 2. The valuable fiber (bast) is in the outer layer of the stem beneath the covering of thin, green tissue.

Cannabis is known to have been used in the Bylony culture of Central Europe (about 7,000 years ago).[184]

The clothes the people wore, the nets they fished and hunted with, and the ropes they used in the earliest machines were all made of the long, strong, and durable fiber, hemp. This valuable fiber separates from the stem of *Cannabis* when the stem decays (rets).

In the early classics of the Chou dynasty, written over 3,000 years ago, mention is often made of "a prehistoric culture based on fishing and hunting, a culture without written language but which kept records by tying knots in ropes. Nets were used for fishing and hunting and the weaving of nets eventually developed into clothmaking."[8] These references may well be to the Yang-shao people.

As their culture advanced, these prehistoric people replaced their animal skins with hemp cloth. At first, hemp cloth was worn only by the more prosperous, but when silk became available, hemp clothed the masses.

People in China relied on *Cannabis* for many more products than fiber. *Cannabis* seeds were one of the grains of early China along with rice, barley, millet, and soybeans. The seeds were ground into a meal, or roasted whole, or cooked in porridge. The ancient tombs of China had sacrificial vessels filled with hemp seed and other grains for the afterlife. From prehistoric times there is a continuous record of the importance of hemp seed for food until the first to second century B.C., when the seed had been replaced by more palatable cereal grains.[7] (An interesting note from the Tung-kuan archives (28 A.D.) records

 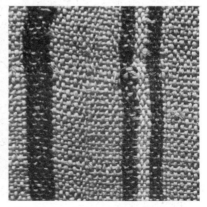

Figures 3 and 4. The making of nets eventually developed into cloth making. (Hemp and contemporary hemp textiles from Japan.)

that after a war-caused famine the people subsisted on "wild" *Cannabis* and soybean.[8])

The effects of *Cannabis'* resinous leaves and flowers did not go unnoticed. The *Pên-ts'ao Ching,* the oldest pharmacopoeia known, states that the fruits (flowering tops) of hemp, "if taken in excess will produce hallucinations" (literally "seeing devils"). The ancient medical work also says, "If taken over a long term, it makes one communicate with spirits and lightens one's body."[9] Marijuana, with a powerful effect on the psyche, must have been considered a magical herb at a time when medical concepts were just being formed. The *Pên-ts'ao Ching,* speaking for the legendary Emperor Shên-nung of about 2000 B.C., prescribes marijuana preparations for "malaria, beriberi, constipation, rheumatic pains, absent-mindedness, and female disorders."[15] Even the *Cannabis* root found its place in early medicine. Ground to form a paste, it was applied to relieve the pain of broken bones and surgery.

New uses were discovered for *Cannabis* as Chinese civilization progressed and developed new technologies. The ancient Chinese learned to mill, heat, and then wedge-press *Cannabis* seeds to extract the valuable oil, a technique still used in the western world in the twentieth century. Pressed seeds yielded almost 20 percent oil by weight. *Cannabis* oil, much like linseed oil, could be used for cooking, to fuel lamps, for lubrication, and as the base in paint, varnish, and soap making. After oil extraction, the residue or "hemp cake" still contained about 10 percent oil and 30 percent protein, a nutritious feed for domesticated animals.

Another advancement came with the Chinese invention of paper. Hemp fibers recycled from old rags and fish nets made a paper so durable that some was recently found in graves in the Shensi province that predates 100 B.C.[9] Hemp paper is known for its longevity and resistance to tearing, and is presently used for paper money (Canada) and for fine Bibles.

The ancient Chinese learned to use virtually every part of the *Cannabis* plant: the root for medicine; the stem for textiles, rope and paper making; the leaves and flowers for intoxication and medicine; and the seeds for food and oil. Some of the products fell into disuse only to be rediscovered by other people at other times.

While the Chinese were building their hemp culture, the

cotton cultures of India and the linen (flax) cultures of the Mediterranean began to learn of *Cannabis* through expanding trade and from wandering tribes of Aryans, Mongols, and Scythians who had bordered China since Neolithic times. The Aryans (Indo-Persians) brought *Cannabis* culture to India nearly 4,000 years ago. They worshipped the spirits of plants and animals, and marijuana played an active role in their rituals. In China, with the strong influence of philosophic and moralistic religions, use of marijuana all but disappeared. But in India, the Aryan religion grew through oral tradition, until it was recorded in the four *Vedas,* compiled between 1400 and 1000 B.C. In that tradition, unlike the Chinese, marijuana was sacred, and the bhangas spirit was appealed to "for freedom of distress" and as a "reliever of anxiety" (from the Atharva Veda).[1] A gift from the gods, according to Indian mythology, the magical *Cannabis* "lowered fevers, fostered sleep, relieved dysentery, and cured sundry other ills; it also stimulated the appetite, prolonged life, quickened the mind, and improved the judgement."[15]

The Scythians brought *Cannabis* to Europe via a northern route where remnants of their campsites, from the Altai Mountains to Germany, date back 2,800 years. Seafaring Europe never smoked marijuana extensively, but hemp fiber became a major crop in the history of almost every European country. Pollen analysis dates the cultivation of *Cannabis* to 400 B.C. in Norway, 150 A.D. in Sweden, and 400 A.D. in Germany and England,[3] although it is believed the plant was cultivated in the British Isles several centuries earlier.[2] The Greeks and Romans used hemp for rope and sail but imported the fiber from Sicily and Gaul. And it has been said that "Caesar invaded Gaul in order to tie up the Roman Empire," an allusion to the Romans' need for hemp.

Marijuana, from its stronghold in India, moved westward through Persia, Assyria and Arabia by 500 A.D. With the rising power of Islam, marijuana flourished in a popular form as hashish. In 1378, the Emir Soudon Sheikhouni tried to end the use of Indian hashish by destroying all such plants, and imprisoning all users (first removing their teeth for good measure). Yet in a few years marijuana consumption had increased.[1]

Islam had a strong influence on the use of marijuana in Africa. However, its use is so ingrained in some ancient cultures of the Zambezi Valley that is appearance clearly predated Islam.

Tribes from the Congo, East Africa, Lake Victoria, and South Africa smoke marijuana in ritual and leisure. The ancient Riamba cult is still practiced in the Congo. According to the Riamba beliefs, marijuana is a god, protector from physical and spiritual harm. Throughout Africa treaties and business transactions are sealed with a puff of smoke from a yard-long pipe.[15]

With increased travel and trade, *Cannabis* seed was brought to all parts of the known world by ships and caravans rigged with the fiber of its kind. And when the first settlers came to the Americas, they brought the seed with them.

CANNABIS AND AMERICAN HISTORY

Like their European forebears, Americans cultivated *Cannabis* primarily for hemp fiber. Hemp seed was planted in Chile in 1545,[64] Canada in 1606, Virginia in 1611, and in the Puritan settlements in Massachusetts in the 1630s.[15] Hemp-fiber production was especially important to the embryonic colonies for homespun cloth and for ship rigging. In 1637, the General Court at Hartford ordered that "every family within this plantation shall procure and plant this present year one spoonful of English hemp seed in some soyle."[12]

Hemp growing was encouraged by the British parliament to meet the need for fiber to rig the British fleets. Partly to dissuade the colonists from growing only tobacco, bounties were paid for hemp and manuals on hemp cultivation were distributed. In 1762, the state of Virginia rewarded hemp growers and "imposed penalties upon those that did not produce it."[2]

The hemp industry started in Kentucky in 1775 and in Missouri some 50 years later. By 1860, hemp production in Kentucky alone exceeded 40,000 tons and the industry was second only to cotton in the South. The Civil War disrupted production and the industry never recovered, despite several attempts by the United States Department of Agriculture to stimulate cultivation by importing Chinese and Italian hemp seed to Illinois, Nebraska, and California. Competition from imported jute and "hemp" (*Musa textilas*) kept domestic production under 10,000 tons per year. In the early 1900s, a last effort by the USDA failed to offset the economic difficulties of a labor shortage and the lack of development of modern machinery for the hemp industry.[64] However, it was legal force that would bring an end to U.S. hemp production.

For thousands of years marijuana had been valued and respected for its medicinal and euphoric properties. The Encyclopaedia Brittanica of 1894 estimated that 300 million people, mostly from Eastern countries, were regular marijuana users. Millions more in both the East and the West received prescription marijuana for such wide-ranging ills as hydrophobia and tetanus.

By the turn of the century, many doctors had dropped marijuana from their pharmacopoeias: drugs such as aspirin, though less safe (marijuana has never killed anyone), were more convenient, more predictable, and more specific to the condition being treated. Pill-popping would become an American institution.

Marijuana was not a legal issue in the United States until the turn of the twentieth century. Few Americans smoked marijuana, and those that did were mostly minority groups. According to author Michael Aldrich,[1] "The illegalization of *Cannabis* came about because of *who was using it*"—Mexican laborers, southern blacks, and the newly subjugated Filipinos.

In states where there were large non-white populations, racist politicians created the myths that marijuana caused insanity, lust, violence and crime. One joint and you were addicted, and marijuana led the way to the use of "equivalent drugs"—cocaine, opium and heroin. These myths were promoted by ignorant politicians and journalists, who had neither experience nor knowledge of *Cannabis,* and grew into an anti-marijuana hysteria by the next generation.

For example, the first states to pass restrictions on marijuana use were in the Southwest, where there were large populations of migrant workers from Mexico. One of the first states to act was California, which, "with its huge Chicano population and opium smoking Chinatowns, labeled marijuana 'poison' in 1907, prohibited its possession unless prescribed by a physician in 1915, and included it among hard narcotics, morphine and cocaine in 1929."[1]

In marijuana, the mainstream society found a defenseless scapegoat to cover the ills of poverty, racism, and cultural prejudice. San Franciscans "were frightened by the 'large influx of Hindoos . . . demanding *Cannabis indica*' who were initiating 'the whites into their habit.' "[11] Editorialists heightened public fears with nightmarish headlines of the "marijuana menace" and "killer weed," and fear of *Cannabis* gradually spread through the West. By 1929, 16 western states had passed punitive restrictions governing marijuana use.

(Sample--Warning card to be placed in R. R. Trains, Buses, Street Cars, etc.)

Beware! **Young and Old — People in All Walks of Life!**

This **may be handed you**

by the friendly stranger. It contains the Killer Drug "Marihuana"-- a powerful narcotic in which lurks
                    Murder!   Insanity!   Death!
                    WARNING!
                    Dope peddlers are shrewd! They may
                    put some of this drug in the 🫖 or
                    in the cocktail **or in the tobacco cigarette.**
          WRITE FOR DETAILED INFORMATION, ENCLOSING 12 CENTS IN POSTAGE — MAILING COST
Address: THE INTER-STATE NARCOTIC ASSOCIATION
                    (Incorporated not for profit)
53 W Jackson Blvd.   Chicago, Illinois, U. S. A.

Figure 5.

Marijuana was not singled out by anti-drug campaigners. During this time, Congress not only banned "hard" narcotics, but also had prohibited alcohol and considered the prohibition of medicinal pain killers and even caffeine.

The Federal Bureau of Narcotics was established in 1930 with Harry Anslinger as its first commissioner. During the first few years of operation, the bureau minimized the marijuana problem, limited mostly to the Southwest and certain ghettos in the big cities of the East. However, the bureau was besieged with pleas from local police and sheriffs to help with marijuana problems. The FBN continued to resist this pressure, because Commissioner Anslinger had serious doubts as to whether federal law restricting marijuana use could be sustained as constitutional. Further, FBN reports indicate that the bureau did not believe that the marijuana problem was as great as its public reputation. Control of the drug would also prove extremely difficult, for as Anslinger pointed out, the plant grew "like dandelions."[11]

The joblessness and misery of the depression added impetus to the anti-marijuana campaign. This came about indirectly, by way of focusing public sentiment against migrant and minority workers who were blamed for taking "American" jobs. Much of this sentiment grew out of cultural and racial prejudice and was supported by groups such as Key Men of America and the American Coalition. The goal of these groups was to "Keep America American."

However, by 1935 almost every state had restricted marijuana use, and local police and influential politicians had managed to pressure the FBN to seek a federal marijuana law. The constitutional question could be circumvented by cleverly tying restrictions to a transfer tax, effectively giving the federal government legal control of marijuana.

With this new tack, the FBN prepared for congressional hearings on the Marijuana Tax Act so that passage of the bill would be assured. Anslinger and politicians seeking to gain from this highly emotional issue railroaded the Marijuana Tax Act through the 1937 Congress. Anslinger made sure that "the only information that they (the congressmen) had was what we would give them at the hearings."[11] No users were allowed to testify in pot's defense, and doctors and scientists were ridiculed for raising contrary views.[16] The new federal law made both raising and use of the plant illegal without the purchase of a hard-to-acquire federal stamp. The FBN immediately intensified the propaganda campaign against marijuana and for the next generation, the propaganda continued unchallenged.

The marijuana hysteria also ended any hopes for a recovery of the hemp industry. What had been needed was a machine that would solve the age-old problem of separating the fiber from the plant stem, an effort which required considerable skilled labor. The machine that could have revolutionized hemp production was introduced to the American public in the February 1938 edition of *Popular Mechanics*. But the Marijuana Tax Act had been passed four months earlier, and the official attitude toward all *Cannabis* is best illustrated by this quote from Harry J. Anslinger, commissioner of the Federal Bureau of Narcotics: "Now this (hemp) is the finest fiber known to man-

kind, my God, if you ever have a shirt made of it, your grand-
children would never wear it out. You take Polish families.
We'd go in and start to tear it up and the man came out with his
shotgun yelling, 'These are my clothes for next winter!' "[2]

During the war years, after the Japanese had cut off America's supply of manila hemp, worried officials supplied hemp
seed and growing information to Midwestern farmers. In Minnesota, Iowa, Illinois, and Wisconsin, hemp farmers showed their
wartime spirit by producing over 63,000 tons of hemp fiber in
1943.

Unlike many of our ancient domesticated plants, *Cannabis*
never lost its colonizing tendencies or ability to survive without
human help. *Cannabis* readily "escapes" cultivated fields and
may flourish long after its cultivation is abandoned. However,
Cannabis always keeps in contact by flourishing in our waste
areas—our vacant fields and lots, along roads and drainage ditches,
and in our rubbish and garbage heaps. Perhaps it awaits dis-
covery by future generations. The cycle has been repeated many
times.

States that once supported hemp industry are now dotted
with stands of escaped weedy hemp. Weedy hemp grows across

Figure 6. A weedy hemp stand in Nebraska. *Photo by A. Karger.*

the country, except in the Southwest and parts of the Southeast. Distribution is centered heavily in the Midwest. Most of these plants are descended from Chinese and European hemp strains that were bred in Kentucky and then grown in midwestern states during World War II. But some weed patches, such as in Kentucky and Missouri, may go back a hundred years, and patches in New England go back perhaps to revolutionary times.

The Anslinger crusades that continued through the sixties are a fine example of government propaganda and control of individual lives and beliefs. We still feel the ramifications in our present laws and in the fear-response to marijuana harbored by many people who grew up with Anslingian concepts. Poor *Cannabis*, portrayed as a dangerous narcotic that would bring purgatory upon anyone who took a toke—violence, addiction, lust, insanity—you name it, and marijuana caused it. All it ever did to us was get us stoned . . . things slowed down a bit . . . enough to stop and look around.

Hopefully, we are living in the last years of the era of illegal marijuana and the persecution of this plant. *Cannabis* is truly wondrous, having served human needs for, perhaps, 10,000 years. It deserves renewed attention not only for its chemical properties, but also as an ecologically sensible alternative for synthetic fibers in general and especially wood-pulp paper. May *Cannabis* be vindicated.

CANNABIS: SPECIES OR VARIETIES

The 10,000-year co-evolution of *Cannabis* and humanity has had a profound impact on both plant and humans. *Cannabis* has affected our cultural evolution; we have affected the plant's biological evolution.

From small populations of ancient progenitors, hundreds of varieties or strains of *Cannabis* have evolved. These variations can be traced to human acts, both planned and accidental.

Ancient farmers, knowing that like begets like, selected *Cannabis* for certain characteristics to better suit their needs. With the need for fiber, seeds from plants with longer stems and better fibers were cultivated. Gradually, their descendants became taller, straight-stemmed, and had a minimum of branches.

Some farmers were interested in seed and oil. They developed large-seeded, bushy plants that could bear an abundance of seeds. Marijuana farmers interested in potency selected plants that flowered profusely with heavy resin and strong psychoactive properties.

The subsequent variations in *Cannabis* are striking. In Italy, where hemp fiber supports a major textile and paper industry, some fiber varieties grow 35 feet in a single season. Other Italian varieties may reach only five or six feet in height, but have slender, straight stems that yield a fiber of very fine quality. In Southeast Asia, some marijuana plants grow only four feet or less, yet these are densely foliated and heavy with resin. Other varieties of marijuana grow 15 to 20 feet in a season and yield over a pound of grass per plant.

Breeding plants is a conscious act. The plant's evolution, however, has also been affected by its introduction to lands and climates different from its original home. Whether plants are cultivated or weeds, they must adapt to their environment. Each new country and growing situation presented *Cannabis* with new circumstances and problems for survival. The plants have been so successful at adapting and harmonizing with new environments that they are now considered the most widely distributed of cultivated plants.[45]

In French, *Cannabis* is sometimes called "Le Chanvre troumper" or "tricky hemp," a name coined to describe its highly adaptable nature. The word adaptable actually has two meanings. The first refers to how a population of plants (the genetic pool) adjusts to the local environment over a period of generations. (The population is, in practice, each batch of seeds you have, or each existing stand or field.) For instance, a garden with some plants that flower late in the season will not have time to seed in the north. The next year's crop will come only from any early seeding plants. Most of them will be like their parents and will set seed early. (See Chapter 18.)

Adaptable is a term that also applies to the individual living plant (phenotype) and, in practical terms, means that *Cannabis* is tenacious and hardy—a survivor among plants. It thrives under a variety of environmental conditions, whether at 10,000 feet in the Himalayas, the tropical valleys of Colombia, or the cool and rainy New England coast.

Through breeding and natural selection, *Cannabis* has evolved in many directions. Botanically and historically, the genus is so diverse that many growers are confused by the mythology, exotic names, and seeming contradictions that surround the plants. Many inconsistencies are explained by understanding how variable *Cannabis* is. There are hundreds of wild, weedy, and cultivated varieties. Cultivated varieties may be useful for only hemp, oil, or marijuana. "Strains," "varieties," "cultivars," "chemovars," or "ecotypes" differ widely in almost every apparent characteristic. Varieties range from two to 35 feet tall; branching patterns run from dense to quite loose, long (five or six feet) or short (a few inches). Various branching patterns form the plant into shapes ranging from cylindrical, to conical, to ovoid, to very sparse and gangly. The shape and color of leaves, stems, seeds, and flowering clusters are all variable characteristics that differ among varieties. Life cycles may be as short as three months, or the plants may hang on to life for several years. Most importantly, different varieties provide great variations in the quality and quantity of resin they produce, and hence in their psychoactive properties and value as marijuana.

The taxonomy (ordering and naming) of *Cannabis* has never been adequately carried out. Early research placed the genus *Cannabis* within the Families of either the Moraceae (mulberry) or the Urticaceae (nettle). Now there is general agreement that the plant belongs in a separate Family, the Cannabaceae, along with one other genus, *Humulus*, the hops plant. (See section on Grafting in Chapter 18.)

A modern scheme for the phylogeny of *Cannabis* would be:
Subdivision . . Angiospermae (flowering plants)
Class . . . . Dicotyledoneae (dicots)
Order . . . . Uriticales (nettle order)
Family. . . . . . Cannabaceae (hemp family)
Genus. . . . . . . *Cannabis* (hemp plant)

Below the genus level, there is no general agreement on how many species should be recognized within *Cannabis*. The *Cannabis* lineage has not been possible to trace after thousands of years of human intervention.

Most research refers to *Cannabis* as a single species—*Cannabis sativa* L. (The word *Cannabis* comes from ancient vernacular names for hemp, such as the Greek *Kannabis; sativa* means "cultivated" in Latin; L. stands for Linnaeus, the botanical author of the name.) But some botanists who are studing *Cannabis* believe there are more than one species within the genus.

Richard Schultes, for example, describes three separate species (see Box A) based on variations in characteristics believed not to be selected for by humans (natural variations) such as seed color and abscission layer (scar tissue on the seed which indicates how it was attached to the stalk).

BOX A

Schultes' Key as it appears in Harvard Botanical Museum Leaflets[45]

1. Plants usually tall (five to 18 feet), laxly branched; akenes* smooth, usually lacking marbled pattern on outer coat, firmly attached to stalk and without definite articulation.

    Cannabis sativa

1A. Plants usually small (four feet or less), not laxly branched; akenes usually strongly marbled on outer coat, with a definite abscission layer, dropping off at maturity.

2. Plants very densely branched, more or less conical, usually four feet tall or less; abscission layer a simple articulation at base of akene.

    Cannabis indica

2A. Plants not branched or very sparsely so, usually one to two feet at maturity. Abscission layer forms a fleshy carbuncle-like growth at base of akene.

    *Cannabis ruderalis***

---

*Akene (or Achene) is the botanical name for the fruit of *Cannabis*. In *Cannabis*, the fruit is essentially the seed.
**Limited to parts of Asia.

Ideally, the classification of living things follows a natural order, reflecting relationships as they occur in nature. Species are groups of organisms that are evolving as distinct units. Biologically, the evolutionary unit is the population, a population being a group of freely inbreeding organisms. Living things don't always fit neatly into scientific categories. And the meaning of species changes with our understanding of life and the evolutionary processes. Often, the definition of species will depend on the particular organism being studied.

A traditional way of defining separate species is that offspring that result from a cross cannot reproduce successfully. As far as is known, all *Cannabis* plants can cross freely, resulting in fully fertile hybrids.[107] But growth habit and actual gene exchange are important considerations in plant taxonomy. If different populations never come in contact, then there is no pressure for them to develop biological processes to prevent mixing. *Cannabis* is pollinated by the wind. Although wind may carry pollen grains hundreds of miles, almost all pollen falls within a few feet of the parent plant. The chance of a pollen grain fertilizing a tiny female flower more than 100 yards away is extremely small.[201] Hence, separate stands or fields of *Cannabis* (populations) are quite naturally isolated. For *Cannabis,* the fact that populations are isolated by distance is not sufficient grounds for labeling them separate species, nor is successful hybridization reason enough to group all populations as one species.

The species question and *Cannabis* mythology are complicated by the plant's ability to rapidly change form and growth habits. These changes can be measured in years and decades, rather than centuries or millennia.

The fact that a pollen grain *does* occasionally fertilize a distant flower leads to a process called introgression. Introgression means that new genes (new variations and possible variations) are incorporated into the population via the foreign pollen. This crossing between populations leads to an increase in variation within the population, but a decrease in the differences between the populations. Although introgression confuses the species question, it also adds to the plant's adaptable nature by providing a resource for adaptive variations. In other words, *Cannabis* has been around. The plants have a rich and varied history of experience, which is reflected in their variety and adaptive nature.

If breeding barriers do not exist, species are often delimited by *natural* differences in morphology (structure or appearance). The natural variations on which Schultes' key is based are actually affected by contact with farmers. For instance, seeds which drop freely from the plant are less likely to be collected and sown by the farmer, so that cultivated *Cannabis* may eventually develop a different type of abscission layer than when wild or weedy.

Seed color and pattern are affected naturally by the need for camouflage. Under cultivation this natural selection pressure would not be the same. Many farmers select seeds by color, believing the darkest are the best developed. In other words, there are serious problems with this limited approach to categorizing species in *Cannabis.* This does not go unrecognized by Dr. Schultes, and the key represents a starting point. However, species should represent distinct groups within a genus, and populations with intermediate characteristics should be the exception. When you grow marijuana, you'll find that most varieties do not fit into any of these categories, but lie somewhere between. The majority of the marijuana from the Western Hemisphere would follow this description: plants tall (eight to 18 feet); well-branched; akenes usually strongly marbled; base of the seed sometimes slightly articulated.

Other characteristics, such as variations on wood anatomy[17] and leaf form,[28] have been suggested for delimiting *Cannabis* species. However, wood anatomy, like stem anatomy, can be seriously affected by selection for hemp in particular, but also by selection for marijuana and seed. Wood anatomy also depends on the portion of the stem examined and on the arrangement of leaves (phyllotaxy), which, in turn, is influenced by light levels, photoperiod, and the physiological development of the plant.

Most *Cannabis* plants have compound leaves with seven to nine blades or leaflets per leaf. Occasionally, varieties are seen where all the leaves have only one to three blades (monophyllous). Such plants sometimes arise from varieties with compound leaves. The factor is genetic, but carries little weight for the separation of species.

Human selection for particular traits can powerfully alter plants. Six vegetables—cabbage, cauliflower, brussel sprouts, broccoli, kale, and kohlrabi—are all descended from a single wild species of mustard herb, *Brassica oleracea.*[216] Human preference for particular parts of the plant led to their development. All six are still considered one species.

Any classification of species in *Cannabis,* based solely on morphological grounds, will prove difficult to justify with our present knowledge of the plant. At this time it seems that all *Cannabis* should be considered one species, *Cannabis sativa* L.

Figure 7. Common marijuana leaf with seven blades (Colombian).

Figure 8. Four leaf types from Colombian marijuana varieties. (Another leaf appears on Plate 15.)

Figure 9. Leaf blades from Figure 8.

The debate on whether there is more than one species has been intense, for the issue has legal implications. Many laws specifically prohibit only *Cannabis sativa.* Presumably other species would not be prohibited. However, in the United States, this argument was recently dismissed when tested in a California court. The court upheld the argument that the law's intent is clear, although it may be questionable botanically: under law all *Cannabis* are regarded alike.

Luckily, the controversy over the number of species is of no more than academic interest to the marijuana grower. The most important characteristic to enthusiasts is the quality or potency of the grass they'll grow.

Potency is mostly a factor of heredity. *The quality of the grass you grow depends on how good its parents were, so choose seeds from the grass you like best.*

The environment has an impact, too, but it can only work on what is contained in the seed. A potent harvest depends on an environment which encourages the seed to develop to a full and potent maturity. The way to begin is to find the most potent grass you can; then you will have taken the first step.

Chapter Two
Cannabinoids: The Active
Ingredients of Marijuana

Cannabis is unique in many ways. Of all plants, it is the only genus known to produce chemical substances known as cannabinoids. The cannabinoids are the psychoactive ingredients of marijuana; they are what get you high. By 1975, 37 naturally occurring cannabinoids had been isolated,[121] and more have since been discovered.[115,118] Most of the cannabinoids appear in very small amounts (less than .01 percent of total cannabinoids) and are not considered psychoactive, or else not important to the high. Many are simply homologues or analogues (similar in structure or function) to the few major cannabinoids which are listed.

1. (−)-Δ^9-trans-tetrahydrocannabinol*—This (Δ^9 THC) is the main psychotomimetric (mindbending) ingredient of marijuana. Estimates state that 70 to 100 percent[121] of the marijuana high results from the Δ^9 THC present. It occurs in almost all *Cannabis* in concentrations that vary from traces to about 95 percent of all the cannabinoids in the sample. In very potent varieties, carefully prepared marijuana can have up to 12 percent Δ^9 THC

---

*There are several numbering systems used for cannabinoids. The system in this book is most common in American publications and is based on formal chemical rules for numbering pyran compounds. Another common system is used more by Europeans and is based on a monoterpenoid system which is more useful in considering the biogenesis of the compound.

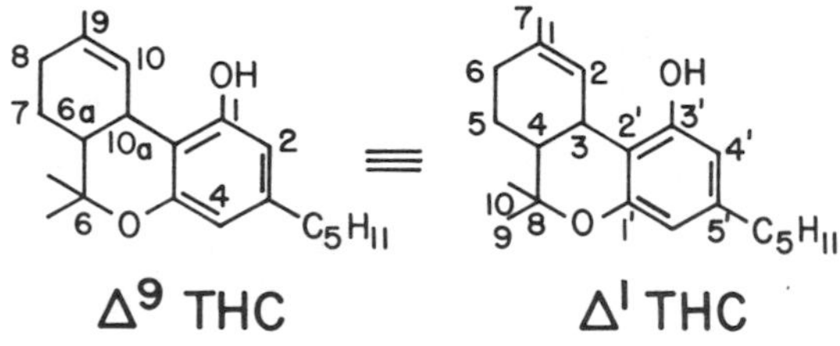

Figure 10.

by dry weight of the sample (seeds and stems removed from flowering buds).*

Δ^8 THC—This substance is reported in low concentrations, less than one percent of the Δ^9 THC present. Its activity is slightly less than that of Δ^9 THC. It may be an artifact of the extraction/analysis process. Here we refer to Δ^9 THC and Δ^8 THC as THC.

2. Cannabidiol—CBD also occurs in almost all varieties. Concentrations range from nil[119,138] to about 95 percent of the total cannabinoids present. THC and CBD are the two most abundant naturally occurring cannabinoids. CBD is not psychotomimetric in the pure form,[192] although it does have sedative, analgesic, and antibiotic properties. In order for CBD to affect the high, THC must be present in quantities ordinarily psychoactive. CBD can contribute to the high by interacting with THC to potentiate (enhance) or antagonize (interfere or lessen) certain qualities of the high. CBD appears to potentiate the depressant effects of THC and antagonize its excitatory effects.[186] CBD also delays the onset of the high[183] but can make it last considerably longer (as much as twice as long). (The grass takes a while to come on but keeps coming on.) Opinions are conflicting as to whether it increases or decreases the intensity of the high,

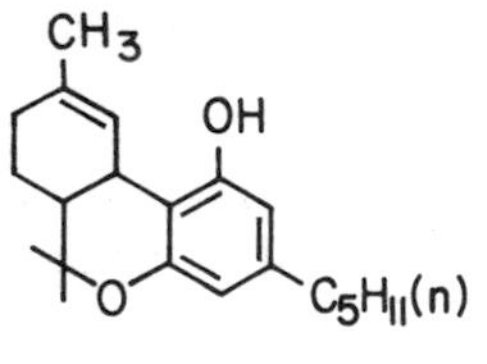

Figure 11. Delta-nine-THC. 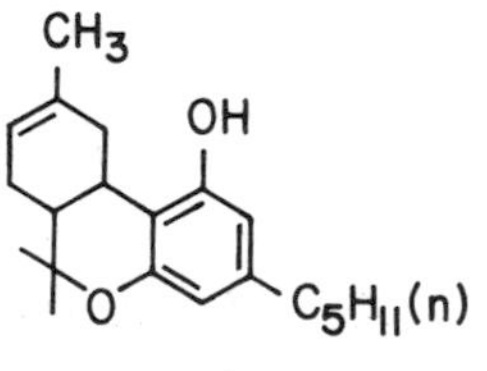   Figure 12. Delta-eight-THC. 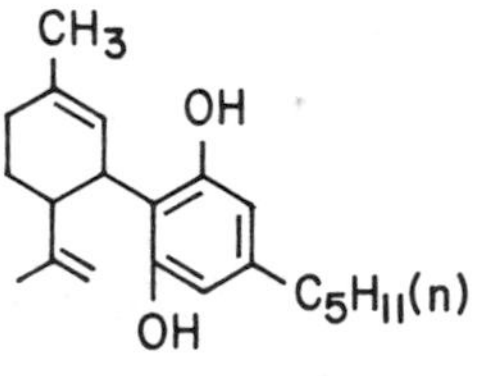   Figure 13. CBD.

"intensity" and "high" being difficult to define. Terms such as knock-out or sleepy, dreamlike, or melancholic are often used to describe the high from grass with sizable proportions of CBD and THC. When only small amounts of THC are present with high proportions of CBD, the high is more of a buzz, and the mind feels dull and the body de-energized.

3. Cannabinol—CBN is not produced by the plant per se. It is the degradation (oxidative) product of THC. Fresh samples of

---

*"Buds" of commercial marijuana is the popular name given to masses of female flowers that form distinct clusters.

marijuana contain very little CBN but curing, poor storage, or processing such as when making hashish, can cause much of the THC to be oxidized to CBN. Pure forms of CBN have at most 10 percent of the psychoactivity of THC.[192] Like CBD, it is suspected of potentiating certain aspects of the high, although so far these effects appear to be slight.[183,185] CBN seems to potentiate THC's disorienting qualities. One may feel more dizzy or drugged or generally untogether but not necessarily higher. In fact, with a high proportion of CBN, the high may start well but feels as if it never quite reaches its peak, and when coming down one feels tired or sleepy. High CBN in homegrown grass is not desirable since it represents a loss of 90 percent of the psychoactivity of its precursor THC.

4. Tetrahydrocannabivarin—THCV is the propyl homologue of THC. In the aromatic ring the usual five-carbon pentyl is replaced by a shorter three-carbon propyl chain. The propyl cannabinoids have so far been found in some varieties originating from Southeast and Central Asia and parts of Africa. What are considered some very potent marijuana varieties contain propyl cannabinoids. In one study, THCV made up to 48.23 percent (Afghanistan strain) and 53.69 percent (South Africa) of the cannabinoids found.[136] We've seen no reports on its activity in

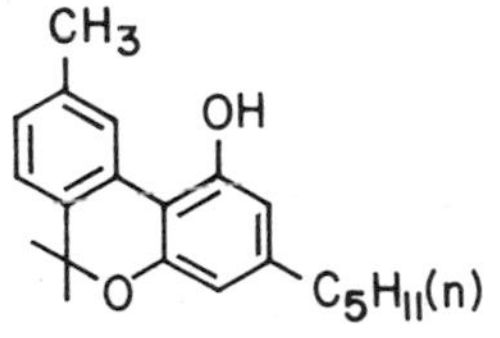

Figure 14. CBN.   Figure 15. THCV.   Figure 16. CBC.

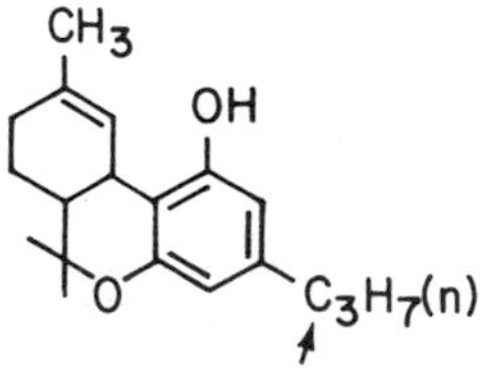

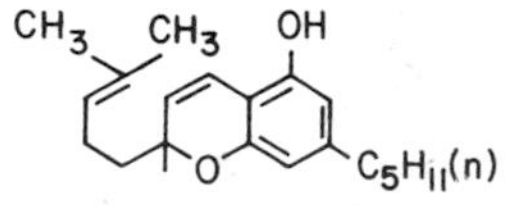

humans. From animal studies it appears to be much faster in onset and quicker to dissipate than THC.[181] It may be the constituent of one- or two-toke grass, but its activity appears to be somewhat less than that of THC.

The propyl cannabinoids are a series corresponding to the usual pentyl cannabinoids. The counterpart of CBD is CBDV; and of CBN, CBV. There are no reports on their activity and for now we can only speculate that they are similar to CBD and CBN. Unless noted otherwise, in this book THC refers collectively to Δ^9 THC, Δ^8 THC, and THCV.

5. **Cannabichromene**—CBC is another major cannabinoid, although it is found in smaller concentrations than CBD and THC. It was previously believed that it was a minor constituent, but more exacting analysis showed that the compound often reported as CBD may actually be CBC.[119,137] However, relative to THC and CBD, its concentration in the plants is low, probably not exceeding 20 percent of total cannabinoids. CBC is believed not to be psychotomimetric in humans[121]; however, its presence in plants purportedly very potent has led to the suspicion that it may be interacting with THC to enhance the high.[137] Cannabicyclol (CBL) is a degradative product like CBN and CBV.[123] During extraction, light converts CBC to CBL. There are no reports on its activity in humans, and it is found in small amounts, if at all, in fresh plant material.

CANNABINOIDS AND THE HIGH

The marijuana high is a complex experience. It involves a wide range of psychical, physical, and emotional responses. The high is a subjective experience based in the individual—one's personality, mood, disposition, and experience with the drug. Given the person, the intensity of the high depends primarily on the amount of THC present in the marijuana. Δ^9 THC is *the* main ingredient of marijuana and must be present in sufficient quantities for a good marijuana high. People who smoke grass that has very little cannabinoids other than Δ^9 THC usually report that the high is very intense. Most people will get high from a joint having Δ^9 THC of .5 percent concentration to material. Grass having a THC concentration of three percent would be considered excellent quality by anyone's standards. In this book, for brevity, we use potency to mean the sum effects of the cannabinoids and the overall high induced.

Marijuana (plant material) is sometimes rated more potent that the content of Δ^9 THC alone would suggest. It also elicits qualitatively different highs. The reasons for this have not been sorted out. Few clinical studies with known combinations of several cannabinoids have been undertaken with human subjects. This field is still in its infancy. So far, different highs and possibly higher potency seem to be due to the interaction of Δ^9 THC and other cannabinoids (THCV, CBD, CBN, and possibly CBC). Except for THCV, in the pure form, these other cannabinoids do not have much psychoactivity.

Another possibility for higher potency is that homologues of Δ^9 THC with longer side chains at C-3 (and higher activity) might be found in certain marijuana varieties. Compounds with longer side chains have been made in laboratories and their activity is sometimes much higher, with estimates over 500 times that of natural Δ^9 THC.[55,113,191] Compounds besides THCV with shorter chains (methyl[139] and butyl[118]) in this position have been found in small amounts in some marijuana samples, indicating that variations do exist. However, this is not a very likely explanation. More likely, THCV is more prevalent in marijuana than supposed and probably had additive or synergistic effects with Δ^9 THC.

The possibility that there are non-cannabinoids that are psychoactive or interacting with the cannabinoids has not been investigated in detail. Non-cannabinoids with biological activity have been isolated from the plants, but only in very small quantities.[181] None are known to be psychotomimetric. However, they may contribute to the overall experience in non-mental ways, such as the stimulation of the appetite.

Different blends of cannabinoids account for highs of different qualities. The intensity of the high depends primarily on the amount of Δ^9 THC present and on the method of ingestion. A complex drug such as marijuana affects the mind and body in many ways. Sorting out what accounts for what response can become quite complex. The methodology to isolate and test the different cannabinoids now exists. The National Institute of Mental Health (NIMH) is funding research on the pharmacology of marijuana. However, such research is paltry, considering that over 30 million people in the United States use the crude drug. Much more research is needed before definite understanding of the cannabinoids and the high is attained.

When the legal restrictions are removed, marijuana will probably be sold by particular blends of cannabinoids and standard amounts of Δ^9 THC. Synthetic marijuana will probably be made with homologues of Δ^9 THC that have much higher activity than the natural form. For now, without access to a lab, you must be satisfied with your own smoking evaluation (for research purposes only), ultimately the most important criterion anyway.

RESIN AND RESIN GLANDS

Many people consider potency and resin concentration synonymous. People hear of plants oozing or gushing with copious resin, and the image is of resin flowing in the plant like the latex of a rubber tree or the sap of a maple tree. But these visions are just pipe dreams. It is quite possible to have a resinous plant with little potency or a plant with little apparent resin which is very potent. Potency depends primarily on the concentration of THC in the plant material. Many more substances besides the cannabinoids make up the crude resin of *Cannabis.* Preparations such as *ghanja* or hashish are roughly about one-third by weight non-psychoactive water-soluble substances and cellular debris. Another third is non-psychoactive resins such as phenoloic and terpenoid polymers, glycerides, and triterpenes. Only one-fourth to one-third is the cannabinoids. In many *Cannabis* plants, THC may be only a very small percentage of the total cannabinoids.* The remainder (5 to 10 percent) of the resin will be essential oils, sterols, fatty acids, and various hydrocarbons common to plants.

The cannabinoids basically do not flow in the plant, nor are they the plant's sap. About 80 to 90 percent of the cannabinoids are synthesized and stored in microscopic resin glands that appear on the outer surfaces of all plant parts except the root and seed. The arrangement and number (concentration) of resin glands vary somewhat with the particular strain examined. Marijuana varieties generally have more resin glands, and they are larger than resin glands on non-drug varieties.

Although resin glands are structurally diverse, they are of three basic types. The bulbous type is the smallest (15-30 μm** or about .0006 to .0012 inches). From one to four cells make up the "foot" and "stalk," and one to four cells make up the "head" of the gland.[25] Head cells secrete a resin—presumably cannabinoids—oils, and related compounds which accumulate between the head cells and the outer membrane (cuticle). When the gland matures, a nipple-like outpocket may form on the

---

*These figures are very approximate. Actual percentages depend on sample material, processing, and extraction procedures. See page 41 for percentages of THC in hashish.
** μm is the symbol for a micrometer (or micron), equal to 1/1,000,000 of a meter, or approximately 1/25,000 of an inch.

membrane from the pressure of the accumulating resin. The bulbous glands are found scattered about the surfaces of the above-ground plant parts.

The second type of gland is much larger and more numerous than the bulbous glands. They are called "capitate," which means having a globular-shaped head. On immature plants, the heads lie flush or appear not to have a stalk and are called "capitate sessile." They actually have a stalk that is one cell high, although it may not be visible beneath the globular head. The head is composed of usually eight, but up to 16 cells, that form a convex rosette. These cells secrete a cannabinoid-rich resin which accumulates between the rosette and its outer membrane. This gives it a spherical shape, and the gland measures from 25 to 100 μm across. In fresh plant material about 80 to 90 percent of their contents will be cannabinoids, the rest primarily essential oils.[146]

During flowering the capitate glands that appear on the newly formed plant parts take on a third form. Some of the glands are raised to a height of 150 to 500 μm when their stalks elongate, possibly due to their greater activity. The stalk is composed mostly of adjacent epidermal tissue. These capitate-stalked glands appear during flowering and form their densest cover on the female flower bracts. They are also highly concentrated on the small leaves that accompany the flowers of fine marijuana varieties. Highest concentration is along the veins of the lower leaf surface, although the glands may also be found on the upper leaf surface on some varieties. The male flowers have stalked glands on the sepals, but they are smaller and less

Figure 17. Upper surface of a small leaf, showing stalked glands (homegrown Colombian, × 16).

concentrated than on the female bracts. Male flowers form a row of very large capitate glands along opposite sides of the anthers.

Capitate-stalked resin glands are the only ones visible without a microscope. To the naked eye, this covering of glands on the female flower bracts looks like talcum or dew sprinkled on a fuzzy surface. With a strong hand lens, the heads and stalks are distinct. Resin glands also can be seen on the anthers of the male flowers and on the undersides of the small leaves that intersperse the flower clusters.

Figure 18. Resin glands on a stem lie close to the surface beneath the cystolith hairs (x40). Hairs always point in direction of growing shoots.

Resin glands are not visible until flowers form. The more obvious covering of white hairs seen on stems, petioles, and leaves are not resin glands. They are cystolith hairs of carbonate and silicate which are common to many plants. These sharp-pointed hairs afford the plant some protection from insects and make it less palatable to larger, plant-eating animals.

In India, to make the finest quality hashish (*nup*), dried plants are thrashed over screens. Gland heads, stalks and trichomes collect in a white to golden powder which is then compressed into hashish (for hashmaking see 87).

Resin rarely accumulates in the copious quantities people would lead you to believe. Actually, the plants form a cover of resin glands rather than a coating of resin. Usually this is no more apparent than for the female flowers to glisten with pin-points of light and for the leaves and stems to feel a bit sticky when you run your fingers over them.

On some fine marijuana strains, resin may become obvious by the end of flowering and seed set. Resins occasionally secrete through pores in the membrane of gland heads. Usually secretion occurs many weeks after the stalked glands appear. The glands seem to empty their contents, leaving hollow spaces (vacuoles) in the stalk and head cells. After secretion, the glands cease to function and begin to degenerate. Gland heads, stalks, and trichomes become clumped together, and the whole flowering surface becomes a sticky mass. For reasons we'll go into later, this is not necessarily desirable. (see Chapters 20, 21.)

Small quantities of cannabinoids are present in the internal tissues of the plant. The bulk is found in small single cells (nonarticulated laticifers) that elongate to form small, individual resin canals. The resin canals *ramify* the developing shoots, and penetrate the plant's conducting tissue (phloem). Minute clumps of resin found in the phloem are probably deposited by these resin canals. Other plant cells contain insignificant amounts of cannabinoids and probably a good 90 percent of the cannabinoids are localized in the resin glands.

Cannabinoid synthesis seems to occur primarily in the head and apex of the stalk cells of the resin glands.[26] Lacticifers and possibly other plant cells probably contribute by synthesizing the simpler molecules that will eventually make up the cannabinoids. Biosynthesis (the way the plant makes the molecules) of the cannabinoids is believed to follow a scheme originally

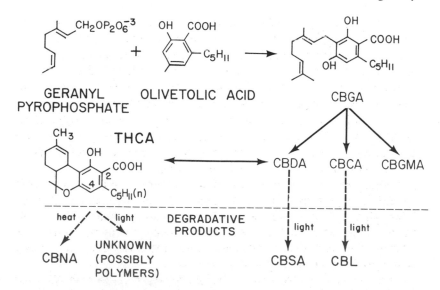

Figure 19. Possible biosynthesis of cannabinoids.

outlined by A.R. Todd in his paper "Hashish," published in 1946 (see Figure 19). In the 1960s the pathway was worked out by Raphael Mechoulam, and confirmed in 1975 by Dr. Shimomura and his associates. Notice that all the cannabinoids are in their acid forms with a (COOH) carboxyl group at C-2 in the aromatic ring. This group may also appear at C-4 and the compounds are called, for example, THC acid "A" and THC acid "B," respectively. The position of the carboxyl group does not affect the potency, but, in fact, in their acid forms the cannabinoids are not psychoactive. In fresh plant material, cannabinoids are almost entirely in their acid forms. The normal procedure of curing and smoking the grass (heat) removes the carboxyl group, forming the gas CO_2 and the psychoactive neutral cannabinoids. Removing the CO_2 is important *only* if you plan to eat the marijuana. It is then necessary to apply heat (baking in brownies, for example) for the cannabinoids to become psychoactive. Ten minutes of baking marijuana at 200°F is enough to convert the THC acids to neutral THC.

The formation of CBG acid, from which all the other cannabinoids are formed, is initially made from much simpler compounds containing terpene units. The example here is olivetolic acid condensing with a terpene moeity called geranyl pyrophosphate. It is not known whether these are the actual or only precursors to CBG in the living plant.

Terpenes and related substances are quite light and some of them can be extracted by steam distillation to yield the "essential oil" of the plant (from essence—giving the flavor, aroma, character). Over 30 of these related oily substances have been identified from *Cannabis*.[143] On exposure to light and air, some of them polymerize, forming resins and tars.

The cannabinoids are odorless; most of the sweet, distinctive, pleasant "minty" fragrance and taste of fresh marijuana comes from only five substances which make up only 5 to 10 percent of the essential oils: the mono- and sesqui-terpenes alpha- and beta-pinene, limonene, myrcene, and beta-phalandrene.[144] These oily substances are volatile and enter the air quickly, dissipating with time. Subsequently, the marijuana loses much of its sweetness and minty bouquet.

The essential oils constitute about .1 to .3 percent of the dry weight of a fresh marijuana sample, or on the order of 10 percent of the weight of the cannabinoids. Essential oils are found within the heads of the resin glands and make up about 10 to 20 percent of their contents in fresh material.[146] They have also been detected in the resin canals (laticifers).[31]

Different samples of *Cannabis* have essential oils of different composition. This is not surprising given the variability of the plant. Since substances found in the essential oils are, or are related to, substances that are the precursors of the cannabinoids, there is some chance that a relationship exists between a particular bouquet and cannabinoid content. No such relationship is yet known, but it has only been studied superficially. When connoisseurs sample the bouquet of a grass sample, they are basically determining whether it is fresh. Fresh grass means fresh cannabinoids and less of these are likely to have been degraded to non-psychoactive products.

PRODUCTION OF CANNABINOIDS BY *CANNABIS*

Why *Cannabis* produces cannabinoids and resins is a question probably every grower has wondered about. Supposedly, if you knew, you could simulate or stimulate an environmental factor to increase cannabinoid production. Unfortunately, it does not follow that increasing a particular selective pressure will affect a plant's (phenotype) cannabinoid production. However, over a period of generations, it is possible that environmental manipulations can increase the overall cannabinoid concentrations in a population of plants. But even this procedure would work slowly compared to direct breeding by the farmer.

From the microstructure of the resin glands and the complexity of the resin, it is apparent that *Cannabis* invests considerable energy in making and storing the cannabinoids. Obviously, the cannabinoids are not a simple by-product or excretory product. No doubt the cannabinoids and resins serve the plant in many ways, but probably they have more to do with *biotic* factors (other living things) rather than *abiotic* factors (non-living environment such as sunlight, moisture, etc.).

The cannabinoids, resins, and related substances make up a complex and biologically highly active group of chemicals, a virtual chemical arsenal from which the plant draws its means for dealing with other organisms. This would apply especially to herbivores, pathogens, and competing plants. In the case of humans, the cannabinoids are an attractant. Some possible advantages to the plant are listed below, but no direct studies have been done on this question. Indeed, it is surprising that botanists have shown so little interest in this question; they have even gone out of their way to state their lack of interest.

Possible Advantages of Cannabinoid Production

1. Obviously the cannabinoids are psychoactive and physiologically active in many animals. This may dissuade plant-eating animals from eating the plant, especially the reproductive parts. Many birds enjoy *Cannabis* seeds (guaranteed to make your canary sing). But in nature, birds will not bother young seeds. probably because they are encased in the cannabinoid-rich bracts. In wild or weedy plants, when the seed is mature it "shells out" and falls to the ground. Birds will eat the naked seeds. However, matured seeds are quite hard. Many will not be cracked and eventually will be dropped elsewhere, helping the plant to propagate. Bees and other insects are attracted to the pollen. The cannabinoids and resins may deter insects from feeding on pollen and developing seeds. Resin glands reach their largest size on the anthers (which hold pollen) and bracts (which contain the seed). See plates 6, 7, 10 and 11.

2. Terpenoid and phenolic resins are known to inhibit germination of some seeds. *Cannabis* resins may help *Cannabis* seedlings compete with other seedlings by inhibiting their germination.

3. Many of the cannabinoids (CBD, CBG, CBC and their acids) are highly active antibiotics against a wide range of bacteria (almost all are gram +).[36,130,184] Crude resin extracts have been shown to be nematocidal.[36] (However, fungicidal activity is low.)

Most of the explanations you've probably heard for resin production from both lore and scientists have to do with physical factors such as sunlight, heat, and dryness. Presumably the resin coats the plant, protecting it from drying out under physical extremes. These explanations make little sense in light of the resins' chemistry.

The physical qualities of the glands and resins probably aid the plant in some ways. The sticky nature of resin may help pollen grains to adhere to the flowering mass and stigmas, or simply make the plant parts less palatable. And gland heads do absorb and reflect considerable sunlight, and so possibly protect the developing seed. For instance, gland heads are at first colorless (i.e., they absorb ultraviolet light). This screening of ultraviolet light, a known mutagen, may lower possible deleterious mutations. But physical properties seem to be secondary to the resins' chemical properties as functional compounds to the plant.

CANNABIS CHEMOTYPES

All *Cannabis* plants produce some cannabinoids. Each strain produces characteristic amounts of particular cannabinoids. Strains differ in the total amounts they contain. Usually they average about three percent cannabinoids to dry weight, but concentrations range from about one to 12 percent cannabinoids in a cleaned (seeds and stems removed), dried bud. Strains also differ in which cannabinoids they produce. Based on *which* cannabinoids, *Cannabis* strains can be divided into five broad chemical groups.* The general trend is for plants to have either THC or CBD as the main cannabinoid.

Type I

Strains are high in THC and low in CBD. This type represents some of the finest marijuana strains. They usually originate from tropical zones below 30 degrees latitude, which in the north runs through Houston and New Orleans to Morocco, North India, and Shanghai, and in the south through Rio de Janeiro, South Africa, and Australia. Most of the high-quality marijuana from Mexico, Jamaica, and Colombia sold in this country is this type; most of you will grow this type. As with all five chemical types, type I comes in different sizes and shapes. Most common are plants about 10 to 12 feet tall (outdoors), quite bushy, with branches that grow outward to form the plant into a cone (Christmas tree shape). Other tall varieties (to 18 feet) have branches that grow upward (poplar-tree shaped—some Mexican, Southeast and Central Asian varieties). A less common short

---

*Chemical classification based on work by Small *et al.*[51]

Table 1.

Type I Plants

COUNTRY OF ORIGIN	SEX[a]	SEED CODE OR SAMPLE NO.	Percentage by Weight[b] of		
			THC	CBD	CBC
Brazil		BR-A	2.16%	t	0.03%
France		FR-C	3.20	0.11%	0.38
Costa Rica		M-173	3.72	0	0.24
India		IN-E	3.31	0.02	0.52
Jamaica	J	JA-B	2.20	0.01	0.41
Mexico		K24-6F	2.65	0	0.15
Mexico		K24-10J	2.97	0	0.16
Thailand		TI-B	2.91	0.42	0.02
South Africa		SA-A(1)	3.02	0.61	0.03
Vietnam	M	VN-A(1)	3.23	0.54	0.23

[a]Sex: M, male; blank, female; J, juvenile or not showing flowers.
[b]Weight of cleaned, dried marijuana samples. t, trace: less than .01 percent of total cannabinoids.

variety (up to eight feet) develops several main stems and the plants appear to sprawl (Mexico, India).

Type II

This is an intermediate group, with high CBD and moderate to high THC. They usually originate from countries bordering 30 degrees latitude, such as Morocco, Afghanistan and Pakistan. In this country, this type of grass usually comes from Afghani and Colombian varieties. Type II plants are quite variable in the intensity and quality of the high they produce, depending on the relative amounts of THC and CBD in the variety. Probably because of their high CBD and overall resin content, these plants are often used to prepare hashish and other concentrated forms of marijuana.

The most common varieties grow to about eight to 12 feet and assume a poplar-tree shape with long branches that grow upward from the stem base and much shorter branches toward the top. They usually come from Turkey, Greece, and Central or Southeast Asia and occasionally from Colombia and Mexico. Some varieties are shorter, about four to eight feet at maturity, and very bushy with a luxuriant covering of leaves. These usually

Figure 20. *Left:* This Pakistani variety ("indica") reaches a height of five feet (large leaves removed). *Right:* Flowering top two months later.

Table 2.

Type II Plants[a]

COUNTRY OF ORIGIN	SEX	SEED CODE	Percentage by Weight of		
			THC	CBD	CBC
Afghanistan	M	AF-B	2.68%	1.94%	0.50%
Afghanistan	M	AF-B(1)	2.63	4.58	0.27
Ethiopia	M	ET-A	1.29	3.05	0.15
France		FR-B	1.49	0.95	0.13
India	J	IN-I	0.99	3.40	0.26
Lebanon		LE-A	1.07	1.68	0.05
Manchuria		MN-A	1.99	1.93	0.23
Thailand		TI-C	1.36	2.40	0.06
Thailand		TI-D	1.56	1.25	0.11
Turkey	M	TU-A(2)	1.59	2.79	0.23

[a]See notes for Table 1.

originate from Nepal, northern India, and other parts of Central Asia as well as North Africa. Other varieties appear remarkably like short (five to seven feet) hemp plants, with straight, slender stems and small, weakly developed branches (Vietnam). A common short variety, less than four feet tall (Lebanon, N. Africa), forms a continuous dense cluster of buds along its short stem. They appear remarkably like the upper half of more common marijuana plants.

Type III

Plants are high in CBD and low in THC. These are often cultivated for hemp fiber or oil seed. Usually they originate from countries north of 30 degrees latitude. As marijuana they yield a low-potency grass and are considered non-drug varieties. If you choose your seeds from potent grass, it will not be this type. An example of these plants are Midwestern weedy hemps which are often collected and sold for low-grade domestic grass. The high CBD content can make you feel drowsy with a mild headache long before you feel high. These plants are very diverse morphologically even when categorized by cultivated types. Hemp plants are usually tall (eight to 20 feet) with an emphasis on stem development and minimal branching. Starting from the base, long, even internodes (stem portion from one set of leaves to the next pair) and opposite phyllotaxy (see p. 50) cover a good portion of the stem. Some varieties form long, sparse

Table 3.

Type III Plants[a]

COUNTRY OF ORIGIN	SEX	SEED CODE	Percentage by Weight of		
			THC	CBD	CBC
France		FR-F	0.07%	1.36%	0
Iowa		IO-A	0.10	1.70	0.02
India	M	IN-B(2)	0.11	2.24	0
Iran	M	IR-A(2)	0.18	1.63	0.03
Morocco		MO-A(1)	0.08	1.61	0
Russia		RU-A(1)	0.10	1.79	0.03
Minnesota			0.20	1.65	n/r[b]
Iowa			0.068	2.70	n/r

[a]See notes to Table 1.
[b]n/r: not reported.

branches only on the upper portion of the stem (many Midwest weeds). Other varieties (Kentucky hemp) are the familiar Christmas-tree shape. Seed varieties are usually short (two to eight feet) and very bushy. Branches on some are short, grow outward and are all of approximately the same length, giving the plant a cylindrical shape. Some of the shorter (two to three feet) seed varieties have undeveloped branches, and almost all of the seeds collect in a massive cluster along the top portion of the stalk. Seed plants are often the most unusual-appearing of *Cannabis* plants, and you won't find them in the United States.

As expected, the figures for average THC in Midwestern weeds are quite low. This is consistent with their reputation for low potency. But the range of THC goes up to 2.37 percent in the Illinois study. This is comparable with some of the higher-quality imported marijuana and is consistent with some people's claims that Midwestern weeds provided them with great highs.

Table 4.

Midwestern Weedy Hemps Type III[a]

	Percentage of THC by Weight		Percentage of CBD by Weight	
	AVG.	RANGE	AVG.	RANGE
Illinois[71]	0.37%	0.05 − 2.37%	1.02%	0.15 − 7.10%
Indiana[80]	0.20	t − 1.50	2.50	t − 6.80
Kansas[b]	n/r	0.01 − 0.49	n/r	0.12 − 1.70

[a]See notes to Table 1.
[b]During flowering. See reference 74.

Type IV

Varieties that produce propyl cannabinoids in significant amounts (over five percent of total cannabinoids) form a fourth group from both type I and II plants. Testing for the propyl cannabinoids has been limited and most reports do not include them. They have been found in plants from South Africa, Nigeria, Afghanistan, India, Pakistan, and Nepal with THCV as high as 53.69 percent of total cannabinoids.[136] They usually have moderate to high levels of both THC and CBD and hence have a complex cannabinoid chemistry. Type IV plants represent some of the world's more exotic marijuana varieties.

A fifth type, based on the production of CBGM, which is not psychoactive, is found in northeastern Asia, including Japan, Korea and China. This type is not relevant to us and will not be mentioned again.

There are many different techniques for sampling, extraction, and estimation of cannabinoids in plant material. To minimize differences among research groups, the above data (except for Midwestern weedy hemps) are taken from studies at the University of Mississippi at Oxford.[66,119,136]

Table 5.

Type IV Plants[a]

Percentage by Weight of	Country of Origin	
	Afghanistan[b]	South Africa[c]
THC	4.41%	0.63%
CBD	0.11	0.06
CBC	0.77	0.01
CBN	0.11	0.17
THCV	0.40	1.69
CBDV	0.63	0.05
CBV	0.06	0.02
CBGM	0.15	—

[a]See notes for Table 1.
[b]Male plant; seed code, AF-A. Data represent 89% of total cannabinoids.[136]
[c]Female plant; seed code, SA-E. Data represent 86% of total cannabinoids.[136]

Unfortunately, some of the best Colombian, Mexican, and Thai varieties are not included in the data. Many of these have not been tested until recently, and the figures are not yet published. Under the system for testing at the University of Mississippi, the highest THC variety reached six to eight percent THC in a bud. These seeds originated from Mexico.

These five chemical types are not distinct entities; that is, each type contains several quite different-appearing varieties. Actually, varieties of different types may look more similar than varieties from the same type. But the ability to produce characteristic amounts of particular cannabinoids is genetically based. This means that each type contains certain genes and gene combinations in common, and in biological terms, the plants are called chemical genotypes.

These types may be from virtually any country simply because of the plant's past and ongoing history of movement. The first three can be found in most countries where *Cannabis* is heavily cultivated, although marijuana plants (types I, II, IV) usually originate from lower latitudes nearer the equator. This may be simply explained in terms of cultural practices. Marijuana traditionally has been cultivated in southerly cultures such as India, Southeast and Central Asia, Africa; and in the West in Mexico, Colombia, Jamaica, and Central American countries. On the other hand, useful characteristics must exist before cultures can put them to use and affect selection. And the characteristic (drug or fiber) must maintain itself within the local environment (see page 267).

Non-drug types (type III) usually originate at higher latitudes with shorter growing seasons. A definite gradation exists for non-drug to drug types, starting in temperate zones and moving toward the equator. The same gradation may be found for the appearance of propyl cannabinoids toward the equator. This doesn't mean that the quality of the grass you grow depends on

Table 6.

Contraband Marijuana

TYPE	Percentage of THC to Dry Weight of Cleaned Sample: Average Range
Colombian	
commercial (brown/gold)	1.0–2.5%
connoisseur (gold/brown)	2.5–4.0
exceptional	4.0–8.0
Mexican	
regular	0.5–1.5
fine	1.5–3.0
top-grade	4.0–8.0
Hawaiian	2.0–7.0
Thai Weed (sticks)	2.0–8.0

whether you live in the north or south, but that over a period of
years and decades, a group of plants may drift toward either
the drug or the non-drug type (either rich in THC or rich in CBD).

The majority of the marijuana sold in the United States has
less than one percent THC; and the bulk of this comes from
Mexican and domestic sources. The highest percentages of THC
in marijuana that we've seen are: Colombian (9.7), Mexican
(13.2), Hawaiian (7.8), and Thai sticks (20.2; however, this is
believed to be adulterated with hash oil). The percentages of
THC reported vary greatly, because they depend on the particular method of sampling and estimation used.

Five samples of Colombian Golds, bought in New York City
and San Francisco for from $30 to $50 (1976) an ounce, averaged 2.59 percent THC and 1.27 percent CBN. The CBN represents an average loss of about one-third of the THC originally
present in the fresh plant by the time it reaches American streets.
This is one advantage that homegrowers have, since their marijuana is fresh. In fresh plant material, less than 10 percent of the
THC will have been converted to CBN, as long as the material is
properly harvested, cured, and stored.

Table 7.

Concentrated Forms of Marijuana[69]

	Percentage by Weight of		
TYPE	THC	CBD	CBN
Turkish extract	1.4%	2.8%	t
Fluid extract (40 years old)	0.43	2.7	5.2%
Red oil	10.0	0.88	3.5
Charas tincture	1.4	3.8	4.0
Hash cake (Greece)	2.1	9.8	3.5

By the time hashish reaches the American market, THC content is usually at the low end of the ranges given here, usually
between 1.5 and 4 percent THC. The darker outer layer of
hashish is caused by deterioration. The inner part will contain
the highest concentration of THC.

The average range for hash oil and red oil is 12 to 25 percent
when it is fresh. It is not uncommon for illicit hash oil to have
more than 60 percent THC. However, light, as well as air, very

Table 8.

Seized Hashish[a]

COUNTRY IN WHICH SEIZED	Range of Percentage of			
	THC		CBD	
Greece	1.0	– 15.8%	1.4	– 11.1%
Nepal	1.5	– 10.9	8.8	– 15.1
Afghanistan	1.7	– 15.0	1.8	– 10.3
Pakistan	2.3	– 8.7	6.8[b]	

[a] Figures compiled from many sources.
[b] Only one figure reported.

rapidly decomposes THC in the oil form (see the section on "Storage" in Chapter 21). You can't tell whether the oil will be wondrous or worthless unless you smoke it.

The preparations listed in Tables 9 and 10 are relatively fresh compared to hashish on the American market. Total cannabinoids make up roughly 25 to 35 percent by weight of hashish and resin preparations. Note that the data in these tables are *relative* concentrations.

The very high figures for CBN in hashish indicate that much of the THC is converted to CBN because of processing and aging. During hashmaking many of the gland heads are broken and the

Table 9.

Relative Percentages of Major Cannabinoids from Hashish & Resin Preparations

COUNTRY	Average Percentage of		
	THC	CBD	CBN
Afghanistan	52.0%	36.0%	12.0%[a]
Burma	15.7	16.3	68.0
Jamaica	77.5	9.1	13.4
Lebanon	32.2	62.5	5.3
Morocco	55.0	34.2	10.8
Nigeria	53.7	9.3	37.0
Pakistan	35.7	48.3	16.1
South Africa	75.6	8.4	16.0

[a] Each row sums to 100%.

THC is exposed to light and air. The figures in these tables are typical of what to expect for relative concentrations of THC in hashish on the American market. Actual concentrations are roughly one-fourth to one-third of these figures.

Obviously, THC percentages for hashish and tinctures are not that high compared to fine marijuana. Hashish in the United States seldom lives up to its reputation. The best buy in terms of the amount of THC for the money is hash oil when it is *high quality* and *fresh*. More often a fine homegrown sinsemilla or sometimes a lightly seeded Colombian is the best investment. (Of course, the best value is always what you grow yourself.)

Table 10.

Relative Percentages of Major Cannabinoids in Hashish from Nepal[126]

HASHISH	Percentage[a] of						THC LOST[b]
	THC	CBD	CBN	THCV	CBDV	CBV	
Sample 1	11.5%	35.9%	22.1%	5.7%	12.5%	12.3%	66%
Sample 2	3.4	41.1	24.8	3.0	11.9	15.8	88
Sample 3	5.5	41.2	30.3	2.3	9.1	11.6	85

[a]Each row in these columns sums to 100%.
[b]Percentage of original THC lost as CBN.

Chapter Three
Before Cultivation Begins

CHOOSING SEEDS

Popular market names of different grades of grass, such as Colombian commercial and Mexican regular, are familiar to growers, but each grade actually may encompass many different varieties. For example, there are Colombian Golds that are similar in most respects, but some varieties grow no taller than six feet. The more common types grow 12 to 15 feet under the same conditions. Some Oaxacan *Cannabis* forms several strong upright branches by maturity, and at a glance may seem to have several stems, yet more often, Oaxacan is conical-shaped and grows about 12 feet.

Most of the fine marijuana sold in this country comes from type I plants with THC as the predominant cannabinoid. Type II plants are less common. You might recognize type II plants by the high. This grass takes longer before its effects are felt, but the high lasts much longer than with other marijuana. Type IV plants are the least common; this marijuana seldom reaches the general American market. This type will get you high after only a few tokes. Type III plants are considered non-drug varieties because they are predominantly CBD with little THC. The effects of CBD are not felt unless it is accompanied by a sizable concentration of THC, such as in type II plants. However, a lot of marijuana from these plants is sold in the United States. Some Mexican and Jamaican regular and much of the low-grade domestic is harvested from type III plants.

You may not be able to tell what type plant you're smoking, but you can tell what you like. Seeds from high-quality marijuana will grow into high-quality marijuana plants. If you like the grass you're smoking, you'll like the grass you grow.

The name of your grass has little to do with potency and may have originated in the mind of some enterprising dealer. Always choose your seeds from what you consider to be the best grass. Don't be swayed by exotic names. If you are not familiar with grass of connoisseur quality, ask someone whose experience you respect for seeds. Smokers tend to save seeds from exceptional grass even if they never plan to plant them.

The origin of your grass, even if you knew it for certain, has little to do with whether it will be dynamite or worthless smoke. In both India[45] and Brazil, hemp is grown which is worthless for marijuana. Likewise, extremely potent marijuana plants grow which are useless for hemp fiber. These plants are sometimes found growing in adjacent fields. Most of the fine-quality marijuana varieties developed in those countries nearer to the equator. How much this had to do with environmental conditions or cultural practices is unknown. In either case, marijuana traffic has been so heavy that fine varieties now grow all over the world. For example, in the United States thousands of people now grow varieties from Mexico. These fine varieties originated in Asia and Africa, and many were brought to Mexican farmers by American dealers during the 1960s. As more farmers grew these new varieties, the quality of Mexican grass imported to the United States improved. Fifteen years ago, virtually no grass was grown in Hawaii. Already people are speaking of varieties such as Maui Wowie and Kona Gold.

The color of the grass does not determine its potency. Marijuana plants are almost always green, the upper surface of the leaves a dark, luxuriant green, and the undersurface a lighter, paler green. Some varieties develop reds and purples along stems and leaf petioles. Occasionally, even the leaves turn red/purple during the last stages of growth (see Plate 6). Grasses termed "Red" more often get their color from the stigmas of the female flowers, which can turn from white to a rust or red color, giving the marijuana buds a distinct reddish tinge. The golds and browns of commercial grasses are determined by the condition of the plant when it was harvested—whether it was healthy (green) or dying (autumn colors). How the plants are harvested, cured, and stored also has a serious effect on color. Commercial grasses from Colombia, Mexico, and Jamaica are often poorly cured and packed. Too much moisture is left in the grass, encouraging microbial decomposition; with warm temperatures, whatever green was left disappears, leaving the more familiar browns and golds. By the time they reach the United States, commercial grasses lose about five to 20 percent of their weight in water loss and often smell moldy or musty.

Color also depends on origin—varieties adapted to tropical or high-altitude areas have less chlorophyll and more accessory pigments, giving the plant their autumn colors (accessory pigments protect the plant from excessive sunlight). Varieties adapted to northern climates, where sunlight is less intense, have more chlorophyll and less accessory pigments. The dying leaves often turn light yellow, grey, or rust. Variations in pigment concentrations are also influenced by local light and particularly the soil conditions under which the plants are grown.

The taste of the smoke—its flavor, aroma, and harshness—also depends more on when the marijuana was harvested and how it was handled after it was grown than on the variety or environmental influences.

You can detect subtle differences in the overall bouquet between freshly picked varieties. The environment probably influences bouquet too, but with most commercial grass the harvesting/storing procedures far outweigh these other, more subtle factors. A musty, harsh-smoking Colombian marijuana can give the mildest, sweetest, homegrown smoke when properly prepared. Don't be influenced by the marijuana's superficial characteristics. Choose seeds from the most potent grass.

Grasses of comparable potency can yield plants of different potencies. This is because fine sinsemillas (homegrown, Hawaiians, Thai weeds, and some Mexicans) are carefully tended and harvested at about peak potency. They are also cured and packed well; so they are fresh when they are distributed in the American market. When you smoke them you are experiencing the plant at about its peak potency. The seeds you plant from this grass will produce plants, at best, of about equal potency. Sometimes they are slightly less simply because of differences in growing conditions. Colombian grasses are not usually harvested at their peak potency. Poor curing and packing also leads to a loss in potency. A significant amount (20 percent and up) of the active cannabinoids (THC, CBD) are converted to much less active cannabinoids (CBN, CBS) or inactive ingredients (polymers—tars, resins, oils, etc.). This is also true of many Mexican and Jamaican grasses that are heavily seeded and poorly handled. Homegrown from this grass can produce plants of higher potency than the original, simply because the homegrown is fresh, and is harvested and cured well so that THC content is at its peak.

When choosing seeds you might consider the following broad generalizations. Mexican, Jamaican (if you can find good Jamaican anymore), and homegrowns, including Hawaiians, often develop quickly and have a better chance of fully maturing in the shorter growing seasons over most of the north and central states. Colombian, African, and Southeast Asian varieties, such as Vietnam and Thai sticks (from Thailand and Japan), more often need a longer season to fully develop. Under natural conditions they seldom flower in the short growing season that covers the northern United States.

For indoor growers, the growing season is all year; so it doesn't matter if plants need longer to develop. Mexican and Jamaican plants usually reach full potency in about six months. Colombian and Southeast Asian varieties may need eight or nine months until they reach their maximum THC or general resin content under indoor conditions.

The grass you choose should have a good stock of mature seeds. Thai weed and fine homegrowns (sinsemillas, which are by definition female flower buds without seeds) may have no seeds at all but more often have a few viable seeds. Most Colombian and Mexican grasses contain between one and two thousand seeds per ounce bag or lid of grass. This may sound like an exaggerated figure, but it's not. Look at the photos in Figure 21 showing the yield from some Michoacan buds. The yield is 40 percent grass (1.22 grams, about three joints), 50 percent seeds (1.56 grams or 120 seeds), and 10 percent stems (.3 grams).

Relative to smoking material, seeds are heavy. Colombian grasses average about 50 percent seeds by weight. A film cannister holds about 1,200 Colombian seeds.

Figure 21. Seeded buds often contain more weight in seeds than grass.

Depending on the variety, healthy mature seeds (which are botanically achene nuts) vary in size between 1/12 and 1/4 inches in length. From any variety, choose seeds that are plump and well-formed with well-developed color. Seed colors range from a buff through a dark brown, and from light grey to almost black colors. Often seeds are mottled with brown or black spots, bars, or lines on a lighter field (see Plate 11). Green or whitish seeds are usually immature and will germinate feebly if at all. Fresh seeds have a waxy glimmer and a hard, intact shell. Shiny, very dark brown or black seeds often mean the contents are fermented and the embryo is dead. Fermented seeds crush easily with finger pressure and are hollow or dusty inside. Seeds that are bruised or crushed are also not viable. This happens to some seeds when grass is compressed or bricked.

Fresh, fully matured *Cannabis* seeds have a high rate of germination; 90 percent or better is typical. It is sometimes helpful to have an idea of how many seeds to expect to germinate. You can tell simply by placing a sample number between wet paper towels which are kept moist. Most of the seeds that germinate do so within a few days of each other. After a week or two, count how many of the original seeds germinated. This gives you a rough idea of what to expect from the seeds when planted.

The viability of seeds gradually declines with time; left in the ground, only 40 percent may germinate next season. Seeds are an ideal prey for many fungi, which are responsible for most of their deterioration. In a warm (70°F or over) and humid atmosphere, fungi rapidly destroy seeds. If kept cool and dry in an airtight container, seeds remain viable much longer. A refrigerator is ideal for storage. Seeds stored this way and left in the buds also maintain high viability for over two years.

CANNABIS LIFE CYCLE

Marijuana plants may belong to any one of a number of varieties which follow somewhat different growth patterns. The following outline describes the more common form of growth. Differences between varieties can be thought of as variations on this standard theme.

Cannabis is an annual plant. A single season completes a

generation, leaving all hope for the future to the seeds. The normal life cycle follows the general pattern described below.

Germination

With winter past, the moisture and warmth of spring stir activity in the embryo. Water is absorbed, and the embryo's tissues swell and grow, splitting the seed along its suture. The radical or embryonic root appears first. Once clear of the seed, the root directs growth downward in response to gravity. Meanwhile, the seed is being lifted upward by growing cells which form the seedling's stem. Now anchored by the roots, and receiving water and nutrients, the embryonic leaves (cotlydons) unfold. They are a pair of small, somewhat oval, simple leaves, now green with chlorophyll to absorb the life-giving light. Germination is complete. The embryo has been reborn and is now a seedling living on the food it produces through photosynthesis. The process of germination is usually completed in three to 10 days.

Figure 22. Germination.

Seedling

The second pair of leaves begins the seedling stage. They are set opposite each other and usually have a single blade. They differ from the embryonic leaves by their larger size, spearhead shape, and serrated margins. With the next pair of leaves that appears, usually each leaf has three blades and is larger still. A basic pattern has been set. Each new set of leaves will be larger, with a higher number of blades per leaf until, depending on variety, they reach their maximum number, often nine or 11. The seedling stage is completed within four to six weeks.

Figure 23. The number of blades per leaf increases during the seedling stage.

Vegetative Growth

This is the period of maximum growth. The plant can grow no faster than the rate that its leaves can produce energy for new growth. Each day more leaf tissue is created, increasing the overall capacity for growth. With excellent growing conditions, *Cannabis* has been known to grow six inches a day, although the rate is more commonly one to two inches. The number of blades on each leaf begins to decline during the middle of the vegetative stage. Then the arrangement of the leaves on the stem (phyllotaxy) changes from the usual opposite to alternate. The internodes (stem space from one pair of leaves to the next, which had been increasing in length) begin to decrease, and the growth appears to be thicker. Branches which appeared in the axils of each set of leaves grow and shape the plant to its characteristic form. The vegetative stage is usually completed in the third to fifth months of growth.

Figure 24. *Left:* Opposite phyllotaxy (leaf arrangement). *Right:* Alternate phyllotaxy.

Preflowering

This is a quiescent period of one to two weeks during which growth slows considerably. The plant is beginning a new program of growth as encoded in its genes. The old system is turned off and the new program begins with the appearance of the first flowers.

Flowering

Cannabis is dioecious: each plant produces either male or female flowers, and is considered either a male or female plant. Male

Figure 25. *Left:* A single male flower (×4). *Photo by R. Harris.* *Right:* Male flower clusters.

plants usually start to flower about one month before the female; however, there is sufficient overlap to ensure pollination. First the upper internodes elongate; in a few days the male flowers appear. The male flowers are quite small, about 1/4 inch, and are pale green, yellow, or red/purple. They develop in dense,

Figure 26. *Left:* A single female flower (×4). *Photo by R. Harris.* *Right:* A female flower cluster (young bud).

drooping clusters (cymes) capable of releasing clouds of pollen dust. Once pollen falls, males lose vigor and soon die.

The female flower consists of two small (1/4 to 1/2 inch long), fuzzy white stigmas raised in a V sign and attached at the base to an ovule which is contained in a tiny green pod. The pod is formed from modified leaves (bracts and bracteoles) which envelop the developing seed. The female flowers develop tightly together to form dense clusters (racemes) or buds, cones, or colas (in this book, buds). The bloom continues until pollen reaches the flowers, fertilizing them and beginning the formation of seeds. Flowering usually lasts about one to two months, but may continue longer when the plants are not pollinated and there is no killing frost.

Seed Set

A fertilized female flower develops a single seed wrapped in the bracts. In thick clusters, they form the seed-filled buds that make up most fine imported marijuana. After pollination, mature, viable seeds take from 10 days to five weeks to develop. When seeds are desired, the plant is harvested when enough seeds have reached full color. For a fully-seeded plant this often takes place when the plant has stopped growth and is, in fact, dying. During flowering and seed set, various colors may appear. All the plant's energy goes to reproduction and the continuance of its kind. Minerals and nutrients flow from the leaves to the seeds, and the chlorophylls that give the plant its green color disintegrate. The golds, browns, and reds which appear are from accessory pigments that formerly had been masked by chlorophyll.

About Plants Generally

Plants use a fundamentally different "life strategy" from animals. Animals are more or less self-contained units that grow and develop to predetermined forms. They use movement and choice of behavior to deal with changing environments. Plants are organized more as open systems—the simple physical characteristics of the environment, such as sunlight, water, and temperature, directly control their growth, form, and life cycles. Once the seed sprouts, the plant is rooted in place and time. Since growth is regulated by the environment, development is in accordance with the plant's immediate surroundings. When a balance is struck, the strategy is a success and life flourishes.

Behavior of a plant is not a matter of choice; it is a fixed response. On a visible level the response more often than not is growth, either a new form of growth, or specialized growth. By directly responding, plants in effect "know," for example, when to sprout, flower, or drop leaves to prepare for winter.

Everyone has seen how a plant turns toward light or can bend upward if its stem is bent down. The plant turns by growing cells of different lengths on opposite sides of the stem. This effect turns or rights the plant. The stimulus in the first case is light, in the second gravity, but essentially the plant responds by specialized growth. It is the same with almost all facets of a plant's life—growth is modified and controlled by the immediate environment. The influence of light, wind, rainfall, etc., interacts with the plant (its genetic make-up or genotype) to produce the individual plant (phenotype).

The life cycle of *Cannabis* is usually complete in four to nine months. The actual time depends on variety, but it is regulated by local growing conditions, specifically the photoperiod (length of day vs. night). *Cannabis* is a long-night (or short-day) plant. When exposed to a period of two weeks of long nights— that is, 13 or more hours of continuous darkness each night—the plants respond by flowering. This has important implications, for it allows the grower to control the life cycle of the plant and adapt it to local growing conditions or unique situations. Since you can control flowering, you control maturation and, hence, the age of the plants at harvest.

PHOTOPERIOD AND FLOWERING

For the marijuana grower the most important plant/environment interaction to understand is the influence of the photoperiod. The photoperiod is the daily number of hours of day (light) vs. night (dark). In nature, long nights signal the plant that winter is coming and that it is time to flower and produce seeds, As long as the day-length is long, the plants continue vegetative growth. If female flowers do appear, there will only be a few. These flowers will not form the characteristic large clusters or buds. If the days are too short, the plants flower too soon, and remain small and underdeveloped.

The plant "senses" the longer nights by a direct interaction with light. A flowering hormone is present during all stages of growth. This hormone is sensitive to light and is rendered inactive by even low levels of light. When the lengths of dark periods are long enough, the hormones increase to a critical level that triggers the reproductive cycle. Vegetative growth ends and flowering begins.

The natural photoperiod changes with the passing of seasons. In the Northern Hemisphere, the length of daylight is longest on June 21. Day-length gradually decreases until it reaches its shortest duration on December 22. The duration of daylight then begins to increase until the cycle is completed the following June 21. Because the Earth is tilted on its axis to the sun, day-length also depends on position (or latitude) on Earth. As one moves closer to the equator, changes in the photoperiod are less drastic over the course of a year. At the equator (0 degrees latitude) day-length lasts about 14 hours on June 21 and 10 hours on December 22. In Maine (about 45 degrees north), day-length varies between about 16 and nine hours. Near the Arctic Circle on June 21 there is no night. On December 22 the whole day is dark. The longer daylength toward the north prevents marijuana from flowering until later in the season. Over most of the northern half of the country, flowering is often so late that development cannot be completed before the onset of cold weather and heavy frosts.

The actual length of day largely depends on local conditions, such as cloud cover, altitude, and terrain. On a flat Midwest plain, the effective length of day is about 45 minutes longer than sunrise to sunset. In practical terms, it is little help to calculate the photoperiod, but it is important to realize how it affects the plants and how you can use it to your advantage.

Cannabis generally needs about two weeks of successive long nights before the first flowers appear. The photoperiod necessary for flowering will vary slightly with (1) the variety, (2) the age of the plant, (3) its sex, and (4) growing conditions.

1. *Cannabis* varieties originating from more northerly climes (short growing seasons) react to as little as nine hours of night. Most of these are hemp and seed varieties that are acclimated to short growing seasons, such as the weedy hemps of Minnesota or southern Canada. Varieties from more southerly latitudes need longer nights with 11 to 13 hours of darkness. Since most

marijuana plants are acclimated to southerly latitudes, they need the longer nights to flower. To be on the safe side, *if you give any Cannabis plant dark periods of 13 or more hours, each night for two weeks, this should be enough to trigger flowering.*
2. The older a plant (the more physiologically developed), the quicker it responds to long nights. Plants five or six months old sometimes form visible flowers after only four long nights. Young marijuana plants (a month or so of age) can take up to four weeks to respond to long nights of 16 hours.
3. Both male and female *Cannabis* are long-night plants. Both will flower when given about two weeks of long nights. The male plant, however, will often flower fully under very long days (18 hours) and short nights (six hours). Males often flower at about the same time they would if they were growing in their original environment. For most marijuana plants this occurs during the third to fifth month.
4. Growing conditions affect flowering in many ways (see Chapter 12). Cool temperatures (about 50°F) slow down the flowering response. Cool temperatures or generally poor growing conditions affect flowering indirectly. Flower development is slower, and more time is needed to reach full bloom. Under adverse conditions, female buds will not develop to full size.

Applications of Photoperiod

The photoperiod is used to manipulate the plants in two basic ways:
1. By giving long dark periods, you can force the plants to flower.
2. By preventing long nights, using artificial light to interrupt the dark period, you can force the plants to continue vegetative growth.

Outdoors

Most marijuana plants cultivated in the United States begin to flower by late August to early October and the plants are harvested from October to November. For farmers in the *South, parts of the Midwest, and West Coast, this presents no problem and no special techniques are needed for normal flowering.*
In much of the North and high-altitude areas, many varieties will not have time to complete flowering before fall frosts. To

force the plants to flower earlier, give them longer night periods. If the plants are in containers, you can simply move them into a darkened area each evening. Plants growing in the ground can be covered with an opaque tarpaulin, black sheet plastic, or double- or triple-layered black plastic trash bags. Take advantage of any natural shading because direct sunlight is difficult to screen completely. For instance, if the plants are naturally shaded in the morning hours, cover the plants each evening or night. The next morning you uncover the plants at about eight to nine o'clock. Continue the treatment each day until all the plants are showing flowers. This usually takes two weeks at most, if the plants are well developed (about four months old). For this reason, where the season starts late, it is best to start the plants indoors or in cold frames and transplant outdoors when the weather is mild. This in effect lengthens the local growing season and gives the plants another month or two to develop. By the end of August the plants are physiologically ready to flower; they sometimes do with no manipulation of the photoperiod. More often female plants show a few flowers, but the day-length prevents rapid development to large clusters. The plants seem in limbo—caught between vegetative growth and flowering. Three or four successive long-night treatments may be all that is needed to trigger the plants to full flowering. The natural day-length at this time of year will not be long enough to reverse the process, so you can discontinue the treatment when you see that the new growth is predominantly flowers.

In areas where frosts are likely to occur by early October, long-night treatments may be the only way you can harvest good-sized flower clusters. These clusters, or buds, are the most potent plant parts and make up the desired harvest. Forcing the plants to flower early also means development while the weather is warm and the sun is shining strongly. The flower buds will form much faster, larger, and reach their peak potency. A good time to start the treatments is early to middle August. This allows the plants at least four weeks of flowering while the weather is mild.

Another reason you may want to do this is to synchronize the life cycle of the plants with the indigenous vegetation. In the northeast and central states, the growing season ends quite early and much of the local vegetation dies back and changes color. Any marijuana plants stick out like green thumbs, and the crop may get ripped off or busted. Plants treated with long nights during late July will be ready to harvest in September.

Outdoors, growers should always plant several varieties, because some may naturally flower early, even in the northernmost parts of the country. These early-maturing varieties usually come from Mexican, Central Asian, and homegrown sources. By planting several varieties, many of you will be able to find or develop an early-maturing variety after a season or two. This, of course, is an important point, because it eliminates the need for long-night treatments.

Figure 27. Outdoor August planting yields November flowering under natural light.

Preventing Flowers

Manipulation of the photoperiod can also *prevent* the plants from flowering until a desired time. For example, in Hawaii the weather is mild enough to grow winter crops. The normal summer crop is harvested anytime from September to mid-November. The winter crop is generally planted from October to December. Because the winter days are so short, the plants flower almost immediately, usually within two months. The plants are harvested in their third or fourth month and yield about 1/4 the yield of summer plants. A large Hawaiian female can yield a pound of buds. Most of the plant's overall size is reached while it is vegetatively growing. By interrupting the night period with light, you can keep these plants vegetatively growing for another month, yielding plants of about twice the size.

The amount of light needed to prevent flowering is quite small (about .03 foot candles[95] —on a clear night the full moon is about .01 foot candles). However, each plant must be illuminated fully, with the light shining over the whole plant. This might be accomplished with either electric lights or a strong flashlight. The easiest way is to string incandescent bulbs, keeping them on a timer. The lights need be turned on for only a flash at any time during the night period, from about 9:00 p.m. to about 3:00 a.m. This interrupts the long night period. It can be done at any time of darkness that shortens the night period to less than nine hours. Start these night treatments after the first month of growth. Continue the treatment each night or two, until you want the plants to flower.

Indoors

Natural Light   Indoors, the growing season lasts all year. The night period is much easier to control. Sometimes people grow plants in their windows for more than a year without any female flowers ever forming. This is because household lamps are turned on sometime at night, illuminating the plants. Under natural light exclusively, indoor plants flower at about the same time they would outdoors (sometimes a bit sooner because it is warmer indoors or the plants may be shaded). When plants are well developed and you want them to flower, make sure that no household lamps or nearby streetlamps are shining on them. During late fall and winter, the natural day-length is short enough for the plants to flower naturally, if you simply keep off any lights at night that are in the same room as the plants. If you must use light, use the lowest wattage possible, such as a six-watt bulb. (The hormone is also least sensitive to blue light.) Shield the light away from the plants. Or shield the plants from any household light with aluminum foil curtains. Once the flowers are forming clusters, you can discontinue the dark treatments, especially if it is more convenient. However, if it is too soon (when you see only a few random flowers), household lights can reverse the process.

By using natural light, you can grow indoor crops all year. The winter light is weak and the days are short, so it is best to use artificial lights to supplement daylight, as well as to extend the photoperiod. The extra light will increase the growth rate of the plants and hence size and yield. You should allow winter crops to flower during late January or February, using the natural photoperiod to trigger flowering. If you wait until spring, the natural light period will be too long and may prevent flowering.

Artificial Lights   Under artificial light the photoperiod is, of course, any length you wish. The most popular way to grow with artificial lights is the harvest system. Start the plants under long light periods of from 16 to 18 hours daily. After the plants have reached a good size, usually between three and six months, shorten the light cycle to about 12 hours to force flowering.

To decide exactly when to force the plants to flower, let their growth be the determinant. If male plants are showing their flowers, then the females are physiologically ready to flower. Most of the plant's overall height is achieved during vegetative growth. Some varieties, of course, are smaller and grow more slowly than others. Wait until the plants are nearing the limits of the height of the garden or are *at least five* feet tall. This is large enough to support good flower development and return a good yield. If you turn down the light cycle when the plants are young and small, you'll harvest much less grass because the plants simply can't sustain a large number of flowers.

Some growers prefer a continuous growth system, emphasizing leaf growth and a continuous supply of grass. The light cycle is set for 18 to 24 hours a day. This prevents flowering and the plants continue their rapid vegetative growth. Growing shoots and leaves are harvested as used, and plants are removed whenever they lose their vigor and growth has noticeably slowed. New plants are started in their place. In this way, there will be plants at different growth stages, some of which will be in their rapid vegetative growth stage and will be quite potent. Male plants and some females eventually will form flowers, but the famales will not form large clusters. People often use this system when the lights are permanently fixed. Small plants are raised up to the lights on tables or boxes. This garden never shuts down and yields a continuous supply of grass.

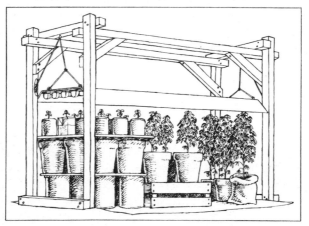

Figure 28. Continuous growth system. *Drawing by O. Williams.*

INHERENT VARIATIONS IN POTENCY

The potency of a particular marijuana sample will vary because of many factors other than the variety. Many of these have to do with the natural development of the plants and their resin glands. Environmental factors do affect potency but there are large differences inherent in any variety. These inherent factors must be explained before we can talk of factors outside the plant that affect relative potency. Strictly environmental effects are discussed in Chapter 19.

Variations in Potency Within Varieties

There are noticeable differences in THC concentrations between plants of the same variety. Differences are large enough so that you can tell (by smoking) that certain plants are better. This is no news to homegrowers, who often find a particular plant to be outstanding. Five-fold differences in THC concentration have also shown up in research. However, when you consider a whole group of plants of the same variety, they're relatively similar in cannabinoid concentrations. Type II plants are the most variable, with individual plants much higher than others in certain cannabinoids.

Variations by Plant Part

The concentration of cannabinoids depends on the plant part, or more specifically, the concentration and development of resin glands to plant part. The female flower bracts have the highest concentration of resin glands and are usually the most potent plant parts. Seeds and roots have no resin glands. These show no more than traces of cannabinoids. Smoking seeds will give you a headache before you can get high. If you got high on seeds, then there were probably enough bracts adhering to the seeds to get you high.

Figure 29. *Left:* The highest concentration of stalked resin glands forms a cover on the female flower bracts (×40). *Right:* Resin glands beneath cystolith hairs on a leaf petiole (×40).

Here are the potencies, in descending order, of the various plant parts:

1. Female flowering clusters. In practice you don't separate hundreds of tiny bracts to make a joint. The whole flowering mass (seeds removed), along with small accompanying leaves, forms the material.

2. Male flower clusters. These vary more in relative potency depending on the strain (see "Potency by Sex," below).

3. Growing shoots. Before the plants flower, the vegetative shoots (tips) of the main stem and branches are the most potent plant parts.

4. Leaves (a) that accompany flowers (small);

             (b) along branches (medium);

             (c) along main stem (large).

Generally, the smaller the leaf is, the more potent it can be.

5. Petioles (leaf stalks). Same order as leaves.
6. Stems. Same order as leaves. The smaller the stem (twig), the higher the possible concentration of cannabinoids. Stems over 1/16" in diameter contain only traces of cannabinoids and are not worth smoking. The small stems that bear the flowers can be quite potent.
7. Seeds and Roots. Contain only traces (less than .01 percent) and are not worth smoking or extracting.

This order is fairly consistent. The exceptions can be the small leaves that accompany male flowers, which are sometimes more potent than the flowers themselves. The growing shoots are sometimes more potent than the mature female flowers, even those found on young plants three or four months old.

Samples of pollen show varying amounts of cannabinoids. Resin glands are found inside the anthers, alongside the developing pollen grains, and form two rows on opposite sides of each anther. Pollen grains are smaller than the heads of large resin glands (see Plate 7), and range from 21 to 69 μm in diameter.[21] A small amount of resin contaminates the pollen when glands rupture, but most of the THC in pollen samples comes from gland heads that fall with the pollen when the flowers are shaken to collect it. One study, using pollen for the standard sample, found concentrations of up to 0.96 percent THC, more than enough to get you high.[79]

Potency by Position on Plant

The potency of marijuana on any plant increases toward the top of the plant, the topmost bud being the most potent. The bottommost leaves on the main stem are the least potent of the usable material. Along branches there is a less steep THC gradient increasing to the growing tip.

The ratios in Table 11 are representative of high-quality marijuana varieties. Plant no. 2 is an exception, with four percent THC in its lower leaves, a figure comparable to high-quality *Colombian* and *Mexican buds* in commercial grass.

Notice the large difference in the gradients of Plants no. 1 and 2, which are from the same variety (SP-5). Like almost all characteristics of these plants, considerable variation occurs even among siblings. Our experience is that generally the better the quality of the variety, the steeper the gradient: in other words, the bigger the difference between top and bottom leaves. For

Table 11.
Relationships of THC Content to Leaf Position[68]

POSITION ON PLANT	Percentage of THC by Weight of Leaf from		
	PLANT NO. 1 (SP-5)	NO. 2 (SP-5)	NO. 3 (UNC-335)
Top	6.1%	6.9%	4.8%
Middle	3.0	5.5	3.1
Bottom	0.8	4.0	1.5
Ratio (gradient)	8:4:1	1.7:1.4:1	3:2:1

example, the plants given here are high-quality type I varieties. Plant no. 1 is more typical, with its steep gradient, than no. 2, where the gradient is much less pronounced. Lower-quality varieties generally do not have as steep a gradient and the ratios would look more like that of Plant no. 2.

Potency by Sex

Although marijuana lore claimed the female to be the more potent, scientists disclaimed this. But there is some truth to both sides. In fine marijuana varieties, male and female leaves average about the same in cannabinoid concentrations. Either a male or a female individual may have the highest concentration in any particular case. The largest variation is in comparing the flowers. Male flowers may be comparable to the females, or they may not even get you high. It seems that the higher the quality of the grass, the better the male flowers will be. In fine type I plants, male flowering clusters usually approach the potency of the female. In low-quality type III varieties, often the male flowers are virtually worthless, with small accompanying leaves being more potent. The leaves of type III females are usually more potent (20 to 30 percent) than the males.

Type II plants are the most variable, with large differences among individual plants. But the trend is for the females to average about 20 percent higher in potency of leaves and flowers.

Potency by Age

In general, the longer the life cycle of the plant, the more the concentration of cannabinoids increases, as long as the plant

Table 12.

Relative Potencies of Male and Female Plants[66]

COUNTRY OF ORIGIN	SEX[a]	Percentage by Weight[b] of		TYPE PLANT
		THC	CBD[c]	
Mexico	M	3.7%	0.86%	I
	F	3.7	0.35	
India	M	4.30	0.12	I
	F	1.78	0.19	
Thailand	M	3.2	0.08	I
	F	3.2	0.42	
India	M	0.81	2.1	II
	F	1.3	0.89	
Pakistan	M	1.37	1.24	II
	F	0.71	1.50	
Turkey	M	0.84	2.11	II
	F	0.92	1.33	
India	M	0.15	2.2	III
	F	0.12	1.2	
Poland	M	0.04	0.97	III
	F	0.06	1.10	

[a]M, male (staminate); F, female (pistillate).
[b]Of flowering mass with accompanying leaves.
[c]Includes CBC.

stays healthy and vigorous. Actually, it is the development of the plant, rather than chronological age, that determines this difference in potency. A plant that is more developed or more mature is generally more potent.

Because you decide when to plant and/or can control the photoperiod, you also control when the plants flower and, hence, the overall age at maturity. A six-month-old plant will generally be better than a four-month-old plant, both of which are flowering. Plants eight months old will usually be more potent than six-month-old plants. Most indoor growers plan their gardens to be about five to eight months old at harvest. Healthy plants can be extended to about 10 months. Plants older than 10 months often develop abnormally. There is usually a decline in vigor and a loss in potency. But some growers have decorative plants several years old.

Outdoor growers more often simply allow the plants to develop according to the local growing conditions which will govern their development and flowering time. Where the growing season is short, some growers start the plants indoors and transplant when the local growing season begins. This gives the plants a longer growing season.

One reason female plants are considered more potent is because of age. Males often flower in four or five months and die, while the females may continue to a ripe old age of eight or nine months, especially when they are not pollinated.

Potency by Growth Stage

Although the general trend is for the cannabinoid concentration to increase with age, this is not a matter of the simple addition or accumulation of cannabinoids. The concentration of cannabinoids changes with the general metabolic rate of the plant, and can be related to the plant's growth pattern or life cycle. Figure 30 shows a hypothetical curve following the concentration of THC from the *upper leaves* and *growing tips* of a male and female plant.

Notice that THC increases immediately with germination and establishment of the seedling, and continues to rise until the plant enters its vegetative stage. At this point, the plant is well-formed, with a sturdy stem, and no longer looks fragile. As the plant's rate of growth increases, there is a corresponding rise in THC that continues throughout the vegetative stage until a plateau is reached. Before the plateau is reached the arrangement of leaves on the stem (phyllotaxy) changes from opposite to alternate. The plateau is maintained until the plant's rapid growth all but stops and the plant has entered preflowering. By this time, the branches have formed the plant to its characteristic shape. Preflowering lasts about one to two weeks, during which THC concentration falls until the appearance of the first flowers.

For the male plant, preflowering ends with renewed growth. This lengthens the uppermost internodes and the first male flower buds appear. THC immediately increases with the development of the male flower clusters, and reaches its peak when most of the flowers are fully formed and a few are beginning to release

66

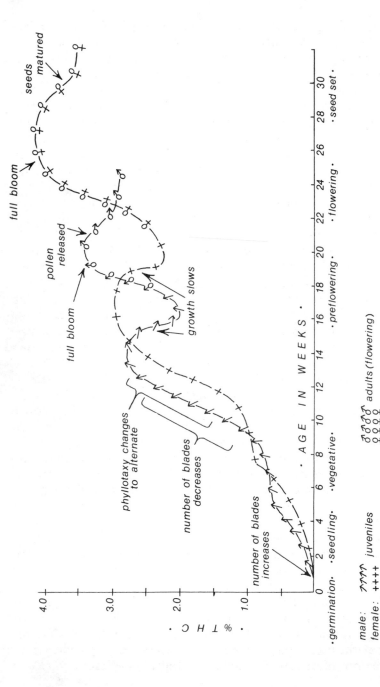

Figure 30. THC graph. *Drawing by E.N. Laincz.*

pollen. After pollen release, the male normally loses vigor and THC content slowly declines until the plant is cured and stored. Female plants reach their maximum THC when the plants are in full bloom. Full bloom is when the plant has filled out with well-formed flower clusters, but flowers are still slowly forming. Most of the stigmas will still be white and healthy.

Flowering lasts anywhere from two to 10 weeks, depending on whether the plants are pollinated or not, as well as on variety and the environment. (See Chapter 20 for details.) THC content declines as the formation of new flowers slows and the majority of the stigmas begin to brown. The only changes you may see in the plants are the maturation of the seeds and the loss of green color in the leaves and flowers. In some cases the plant's apparent resin (its look and feel) increases during the last few weeks of life while the THC concentration is still declining.

You may feel that you should only pick marijuana when the plants are in full bloom, but this is not the case. Think of the garden as a continuous supply of grass. You can never be sure of the fate of your plants. The biggest problem with outdoor growing is that there is a good chance that the plants will be ripped off before you plan to harvest. It is much better to harvest grass during the course of a season, assuring yourself a return for your efforts. For example, during the third month of growth, you could cut back the growing tips, which should be quite potent, often more potent than Figure 30 suggests. This doesn't mean there will be less to harvest at season's end. In fact, the plant will be forced to develop its branches, possibly yielding a larger plant.

Common sense tells you that it is always best to test one sample before you harvest. By taking one tip, curing and smoking, you'll know whether it's worthwhile to harvest more at that time or to wait longer. When a tip is about equal to its parents' potency, then definitely harvest more growing tips. This peak high often occurs during the middle to late rapid, vegetative-growth stage.

The reader should keep in mind that Figure 30 serves only as an example. Chronological age is not as important as the physiological age of the plant. In this graph, the life of the plants is about six months. But the life cycle depends on the particular variety and the growing conditions, which strongly influence the rate of development. (For details on how to use the graph, see Chapter 20.) The important facts that the reader should get from the graph are that the potency of the grass can decrease as well as increase during the plant's life cycle. Actual studies of the cyclic variations in potency over the course of a season have shown much more complicated rhythms, with many more peaks and valleys than here.[71,74,80,86,92] Most varieties will more or less follow a growth pattern as described. Changes in the plant's development, such as phyllotaxy and growth rate, are cues to changes in THC concentration. Secondly, the growing tips of the main stem and branches can be very potent. Growers do not have to wait until flowers form to harvest top-quality smoke.

CULTIVATION: INDOORS OR OUTDOORS?

The basic elements of the environment (light, water, air, and soil) provide plants with their fundamental needs. These environmental factors affect the growth rates of plants, as well as their life cycles. If one factor is deficient, growth rate and vigor will wane regardless of the other three. For instance, with low light, the growth rate will be limited no matter how fertile and moist the soil is. In the same sense, if soil minerals are scarce, the growth rate will be limited no matter how you increase the light.

Photosynthesis

Cannabis, like all green plants, manufactures its food through the process of photosynthesis. Unlike animals, which depend on pre-formed food for survival, plants can use energy from light to form food (carbohydrates) from simple inorganic molecules absorbed from the air and soil.

Plants absorb light energy through pigments that are concentrated in the leaf cells. These pigments are also found in most of the aboveground parts of the plant. The most abundant pigment is chlorophyll, which gives plants their green color. The energy absorbed is stored in chemical compounds such as ATP and $NADPH_2$.* These are storage/transfer compounds that function to transfer energy and matter in the living system. ATP trans-

---

*ATP, adenosine-triphosphate; $NADPH_2$, nicotinamide-adenine-dinucleotide-phosphate.

fers energy that fuels the reactions for the making of carbohydrates as well as most other metabolic functions. $NADPH_2$ transfers electrons, usually as hydrogen, for the synthesis of carbohydrates as well as other compounds.

The raw material for the synthesis of carbohydrates $(CH_2O)^n$ comes from carbon dioxide (CO_2) and water (H_2O). Carbon dioxide is absorbed primarily from the air, but can also be absorbed from the soil. Water is absorbed primarily from the soil and secondarily from the air.

Photosynthesis is summarized as follows:

$$\text{light energy} \xrightarrow{\text{chlorophyll}} ATP + NADPH_2$$

$$CO_2 + H_2O \xrightarrow{ATP, NADPH_2} (CH_2O)_n + O_2$$

For more complex bio-molecules such as amino acids and proteins, the plant absorbs minerals (including nitrogen, phosphorus, and sulfur) from the soil. Carbohydrates provide food energy for the plant using processes similar to those that occur in humans. They also form the basic building blocks for plant tissues. For example, the sugar glucose $(CH_2O)_6$ is strung and bonded to form long chains of cellulose, the most abundant organic compound on earth. About 80 percent of the structure of the plant's cells is made from cellulose.

The plant is a living thing existing in a holistic world; a myriad of factors affect its life. However, good cultivation techniques require attention to only four basic growth factors. With this accomplished, the plants will do the rest.

As grower, your strategy is to bring out the plant's natural qualities. The cannabinoids are natural to the plants. Seeds from potent marijuana grow into potent marijuana plants when they are nurtured to a full and healthy maturity.

Since most marijuana plants are adapted to tropical or semitropical climates, it is up to the grower to make the transition to local growing conditions harmonious. This requires sensible gardening techniques and, in some cases, manipulation of the photoperiod. There is no magic button to push or secret fertilizer to use. The secret of potency lies within the embryo. The environment can and does affect potency, as it does most aspects of the plant's life. However, environmental factors are secondary to the plant's heritage (genetic potential).

Indoors vs. Outdoors

At this point the book divides into separate indoor and outdoor cultivation sections, and you may wonder whether it is better to grow the plants indoors or outdoors. Each alternative has advantages and disadvantages. It is usually better to grow the plants outdoors if possible, because the plants can grow much larger and faster than indoors. Indoors presents space and light limitations. It is possible to grow a 15-foot bush indoors, but this is unrealistic in most homes. There simply isn't enough room or light for such a large plant. Outdoor gardens return a much higher yield for the effort and expense. Most indoor gardeners buy soil and may have to buy electric lights. So there is an initial investment of anywhere from $10 on up.

On the other hand, outdoor plants are more likely to be seen. Many gardens get ripped off, and busts are a constant threat. Indoor gardens are much less likely to be discovered. Gardening indoors allows the grower closer contact with the plants. The plants can be grown all year long; it is an easy matter to control their growth cycles and flowering. Probably the biggest attraction of indoor gardens is that they are beautiful to watch and easy to set up anywhere.

One popular compromise is to construct a simple greenhouse. Use sheet plastic to either enclose part of a porch or to cover a frame built against the house.

The potency of the plants doesn't depend on whether they are grown indoors or outdoors. As long as you grow healthy plants that reach maturity and complete their life cycle, the grass can be as good as any you've ever smoked.

PART 2.
Indoor Gardening

Chapter Four

Introduction

Marijuana adapts well to indoor conditions. You can grow it in sunny rooms or with artificial light. The factor limiting the rate of growth indoors is often the amount of light, since it is less a problem to supply the plants with plenty of water, nutrients, and air.

Natural light is free. If feasible to use, natural light eliminates the most expensive components for indoor gardeners: artificial lights and the electricity they use. Window light is the easiest way to grow plants for decorative purposes or for a small crop. On the other hand, a greenhouse, sunporch, or particularly sunny room can support larger plants than most artificial light systems. A sunny porch or roof area enclosed in sheet plastic to form a greenhouse is a simple, inexpensive way to grow pounds of grass.

Cannabis grows into a fully formed bush when it receives a minimum of five hours of sunlight a day. But you can grow good-sized plants of excellent quality with as little as two hours of daily sunlight provided windows are unobstructed by buildings or trees and allow full daylight. Windows facing south usually get the most light, followed by windows facing east and west (north-facing windows seldom get any sun). Use the location with the longest period of sunlight. The corner of a room or alcoves with windows facing in two or three directions are often very bright. Skylights are another good source of bright, unobstructed light.

Some growers supplement natural light with artificial light from incandescent or fluorescent fixtures. This is essential during the winter, when sunlight is weaker than the summer, and in spaces where the plants get little direct sunlight. Artificial lights can also be used to lengthen the natural photoperiod in order to grow plants all year.

The best time to plant using natural light is in late March or April, when the sun's intensity and the number of hours of daylight are increasing. Cleaning the windows can dramatically increase the amount of light, especially in cities where soot and grime collect quickly. Window sills and sides painted white or covered with aluminum foil reflect light in the plants' direction. Young plants should be placed on shelves, blocks, or tables to bring them up to the light. During warm weather place the plants as close to the window as possible. When the weather is cold, plants should be protected from freezing drafts. You can insulate windows by stapling or tacking clear sheets of polyethylene film to the window frame.

The main problem with growing marijuana in windows is that it may be seen by unfriendly people outside. This won't be a problem for the first couple of months, but when the plants are large they are easily recognized. If this becomes a worry, cover the windows with mesh curtains, rice paper, polyethylene plastic or other translucent materials that will obscure the plants but still let light through. A strip covering the lower part of the window may be enough to interfere with the line of sight from the street.

Many of you will want the garden completely hidden. Some growers opt for closets, basements, attics, spare rooms—even under loft beds. They cover the windows if the garden is visible and grow the plants entirely with artificial light.

The amount of light you provide is what determines the size of your garden—the amount of soil, the number and size of the plants, and the yield of marijuana. Since light is the factor on which you base the planning of your garden, let's begin with artificial light.

Chapter Five
Artificial Light

FIXTURES

Fluorescent light is the most effective and efficient source of artificial light readily available to the home grower. Fluorescent lamps are the long tubes typical of institutional lighting. They require a fixture which contains the lamp sockets and a ballast (transformer) which works on ordinary house current.

Tubes and their fixtures come in lengths from four inches to 12 feet. The most common and suitable are four- and eight-foot lengths. Smaller tubes emit too little light for vigorous growth; longer tubes are unwieldy and hard to find. The growing area must be large enough to accommodate one or more of these fixtures. Overhead space must be sufficient to hang and easily raise the fixtures through a height of *at least* six feet as the plants grow. Fixtures may hold from one to six tubes and may include a reflector, used for directing more light to the plants. Some fixtures are built with holes in the reflectors in order for heat to escape. They are helpful in areas where heat builds up quickly. You can make reflectors with household materials for fixtures not equipped with reflectors. Try to get fixtures that have tubes spaced apart rather than close together. See p. 88 for further suggestions.

The tubes and their appropriate fixtures are available at several different wattages or outputs. Standard or regular output tubes use about 10 watts for each foot of their length—a four-foot tube has about 40 watts and an eight-foot tube about 80 watts.

High Output (HO) tubes use about 50 percent more watts per length than regular output tubes and emit about 40 percent more light. An eight-foot (HO) runs on 112 to 118 watts. Very High Output (VHO) or Super High Output (SHO) tubes emit about two-and-a-half times the light and use nearly three times the electricity (212 to 218 watts per eight-foot tube).

The amount of light you supply and the length of the tube

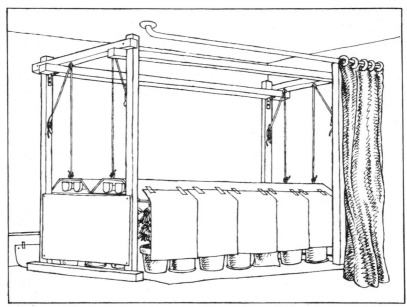

Figure 31. A frame can be constructed with two-by-two's. *Drawing by O. Williams.*

determine the size of the garden. Marijuana will grow with as little as 10 watts per square foot of growing area, but the more light you give the plants, the faster and larger they will grow. We recommend *at least* 20 watts per square foot. The minimum-size garden contains a four-foot fixture with two 40-watt tubes, which use a total of 80 watts. Dividing total watts by 20 (watts per square foot) gives 80W divided by 20W/sq. ft. = four sq. ft. (an area one by four feet). A four-tube (80 watts each) eight-foot fixture would give: 320W divided by 20W/sq. ft. = 16 sq. ft. or an area the length of the tube and about two feet wide.

VHO and HO tubes in practice don't illuminate as wide an area when the plants are young, because the light source is one or two tubes rather than a bank. Once the plants are growing well and the light system is raised higher, they will illuminate a wider area. Figure about 25 w/ft^2 for HO and 35 w/ft^2 for VHO to determine garden size. A two-tube, eight-foot VHO fixture will light an area of 430W divided by 35W/sq. ft. = 12 sq. ft., or an area the length of the tube and one-and-a-half feet wide.

The more light you give the plants, the faster they will grow. Near 50w/sq. ft. a point of diminishing returns is reached, and the yield of the garden is then limited by the space the plants have to grow. For maximum use of electricity and space, about 40w/sq. ft. is the highest advisable. Under this much light the growth rate is incredible. More than one grower has said they can hear the plants growing—the leaves rustle as growth changes their position. In our experience, standard-output tubes can work as well as or better than VHO's if four or more eight-foot tubes are used in the garden.

The yield of the garden is difficult to compute because of all the variables that determine growth rate. A conservative estimate for a well-run garden is one ounce of grass (pure smoking material) per square foot of garden every six months.

In commercial grass, the seeds and stems actually make up more of the bulk weight than the usable marijuana.

The grass will be of several grades depending on when and what plant part you harvest. The rough breakdown might be 1/3 equal to Mexican regular, 1/3 considered real good smoke, and the rest prime quality. With good technique, the overall yield and the yield of prime quality can be increased several fold.

SOURCES

When sunlight is refracted by raindrops, the light is separated according to wavelengths with the characteristic colors forming a rainbow. Similarly, the white light of electric lights consists of all the colors of the visible spectrum. Electric lights differ in the amount of light they generate in each of the color bands. This gives them their characteristic color tone or degree of whiteness.

Plants appear green because they absorb more light near the ends of the visible spectrum (red and blue) and reflect and transmit more light in the middle of the spectrum (green and yellow). The light energy absorbed is used to fuel photosynthesis. Almost any electric light will produce some growth, but for normal development the plants require a combination of red and blue light.

Sunlight has such a high intensity that it can saturate the plants in the blue and red bands, though most of the sun's energy is in the middle of the spectrum. Artificial lights operate at lower intensities; so the best lights for plants are those that emit most of their light in the blue and red bands.

Fluorescent Tubes

Tubes that are manufactured for plant growth generate more of their light in the red and blue bands and are themselves light purple or pink. Gro-Lux (Sylvania), Plant-gro (GE), and Gro-lum (Norelco) appear light purple; Agro-Lite (Westinghouse) and Wide Spectrum Gro-Lux (Sylvania) look more pink. Their spectrum is broader and better-balanced than the standard gro-tubes and more of their energy is in the violet and longer red bands. All of these gro-tubes work well and are popular with growers.

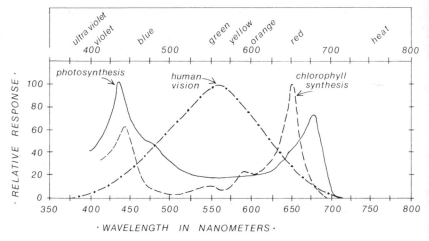

Figure 32. The action spectra of chlorosynthesis and photosynthesis compared to that of human vision. Adapted from *IES Lighting Handbook*. Drawing by E.N. Laincz.

Vita-lite and Optima (Duro-Test) and Natur-escent were originally designed for human vision and produce a white light with a spectrum very similar to natural daylight. These are more expensive but last almost twice as long as other fluorescents, and work about as well as the gro-tubes. They are also available in "power twists" which emit about 10 percent more light than regular tubes and work on regular ballasts.

Standard fluorescent tubes and other lamps used for human vision work almost as well as the above-mentioned ones, although they must be used in combinations to work best. Used singly, they are deficient in either the red or blue band. They are available at any hardware store and are cheaper than gro-tubes. Manufacturers use standardized names such as Daylight or Soft White to designate a tube that has a certain degree of whiteness. Each name corresponds to a tube that emits a particular combination of light in each of the color bands. For example, Cool-White emits more blue light than other colors and appears blue-white. By matching a tube that emits more blue light with a tube with more red light, the bulbs complement each other and produce a healthy growth. It takes more red light to saturate the plants than blue light, so use one blue tube with either two or three red tubes (see Box B).

BOX B

Use one blue fluorescent:
   Blue (decorator)
   Daylight   standard
   Cool White   fluorescents

with two or three red lamps:
   Natural White*
   Merchandizer White
   Soft White   standard
   Warm White Delux fluorescents
   Cool White Delux
   Agro-Lite*
   W/S Gro-Lux*   gro-tubes

---

*These can also be used by themselves as a single source.

Incandescents and Flood Lights

The common screw-in incandescent bulb produces light mainly in the longer wavelengths: far-red, red, orange, and yellow. Higher-wattage bulbs produce a broader spectrum of light than lower-wattage bulbs. Incandescents can be used alone to grow marijuana, but the plants will grow slowly and look scraggly and yellow. Incandescents combined with fluorescents work well, but fluorescents are a better source of red light. Fluorescent tubes generate slightly less heat per watt. With incandescents, heat is concentrated in the small bulb area, rather than the length of the tube, and can burn the plants. In addition, incandescents have less than one-third the efficiency of fluorescents in terms of electricity used. If you decide to use incandescents in combination with fluorescents, use two times the wattage of incandescents to blue source fluorescents, that is, two 40-watt Daylight tubes to about three 60-watt incandescents, evenly spacing the red and blue sources.

The common floodlight has a spectrum similar to but somewhat broader than incandescents. Because they cast their light in one direction and operate at higher intensities, these lights work better than incandescents, both as a single source and to supplement natural or fluorescent light.

The best application for floodlights and incandescents is to

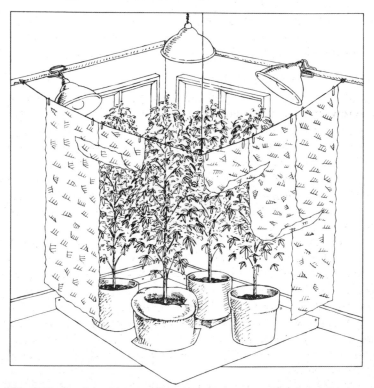

Figure 33. Supplement natural light with floodlights. Use foil curtains for reflectors. *Drawing by O. Williams.*

supplement natural and fluorescent light, especially when the plants get larger and during flowering. Incandescents and floodlights require no special fixtures, although reflectors increase the amount of light the plants receive. These lights are easy to hang or place around the sides of any light system, and their strong red band promotes more growth and good flower development. Some of their energy is in the far-red band. Most purple gro-tubes and white fluorescents are deficient in this band, and addition of a few incandescents make them more effective. Agro-lite and W/S Gro-Lux emit adequate far-red light and need no addition of incandescents.

Several companies make screw-in spotlights specifically for plant growth. Two brand names are Duro-Test and Gro n' Sho. Although they are an improvement over incandescents as a single source, these lights don't perform nearly as well as fluorescents. A 150-watt bulb would grow one plant perhaps four feet tall. Two eight-foot fluorescent tubes (160 watts) will *easily* grow eight six-foot plants. For supplemental lighting, the incandescents and floodlights work as well and are cheaper.

Metal Halide Lamps (HID Lamps)

Metal halide lamps are high-intensity discharge lamps. They have a screw-in base and require a special fixture with a ballast to operate. These, and other high-intensity discharge lamps (sodium and mercury), illuminate factories, streets, and sports complexes and are beginning to be used for horticultural lighting.

We have never grown marijuana with these lamps, but on theoretical grounds, for large installations, they should perform even better than the fluorescents in terms of economy and actual plant growth.

The sizes available are 150, 200, 400, 1,000, and 1,500 watts. Because of their high output from a single bulb, they are easier to install and handle than a large bank of fluorescent lights. For large installations they are cheaper. For supplemental light in greenhouses, the fixture doesn't shade natural light as much as bulky fluorescent fixtures. One distributor told us that marijuana growers are buying one large lamp (1,000 watts) instead of three or four smaller bulbs because of reduced cost and easier handling.

The metal halide lamp has worked the best of any lamp type tested for growing plants. In horticultural studies with several vegetable and grain crops, improvements in yield averaged about

38 percent more dry yield, and ranged from 25 to 100 percent higher when compared with fluorescent or incandescent lights. Among the advantages were longer bulb life, higher output per watt at burnout, about 25 percent more light output per watt, and more flexibility in installation and space.[209]

Metal halide lamps designed for human vision have a broad light spectrum capable of supporting healthy growth. Several manufacturers make color-corrected metal halide lamps specifically for plant growth which produce more light in the red and blue bands. Two are Metalarc (Sylvania) and Multi-Vapor (GE). Metalarc has a particularly well-balanced spectrum for plant growth. Horticultural lamps are designated with HOR in their ordering codes.

General Types of Lamps

The light spectrum (i.e., lamp type) has little to do directly with the potency of the crop. It does affect rate of growth and normal development of the plant. Any of the tubes or lamps, or any of the combinations given, will work well. W/S Gro-Lux is usually cheaper than other gro-tubes and works very well. Our best buy would be a Daylight tube used with every two or three W/S tubes. Daylight with Agro-lite is another good combination.

The biggest difference when comparing tube types is simply what the plants look like under the different colored lamps. The light spectrum will also affect the pigment concentrations and hence color of the leaves. But this is a small and superficial difference. You might consider the color of the lights for decor if the lights are in a room you frequently use. Most stores have fixtures where you can check the tubes to see their color.

The standard fluorescents used for human vision work about as well as the gro-tubes when used in the combinations mentioned. They emit about 10 percent more overall light energy than the gro-tubes. More of this light is in the middle of the spectrum (green, yellow, orange), which is used in photosynthesis to a lesser extent, however, than red and blue light.

The more usable light a plant receives, the faster and larger it will grow. A larger quantity of light increases the *amount* of high-potency grass. It affects potency only marginally on the low end—when the plants are not receiving enough light. A garden with *four eight-foot regular-output tubes gives the best results for a modest investment.* Six tubes work even better.

Some marijuana varieties are capable of growing 12 to 15 feet tall under good outdoor conditions. In practice most indoor plants grow no more than six to eight feet tall because of the amount of light they receive and height limitations of the light system. Height is not that important. Under low light, a plant may grow quite tall but may only produce a small number of leaves and flowers.

The rate and density of growth are most important to yield. Ideally, the plants fill a cubic area with dense foliage. During flowering, a thicket of branch and stem tips laden with thick buds will develop for the final harvest. Figures 34 and 35 show how rapidly the intensity of light diminishes as the distance from the light increases. Even under strong light, the bottom leaves of tall plants often drop from lack of light.

Watts per square foot is a useful measure but not truly accurate in terms of usable output or actual plant growth. With longer tubes or with more light, the growth rate increases and the *cubic* area that can support active growth increases. Hence, plants become fuller and larger. For example, a four-tube (320W) eight-foot garden is twice the area of a two-tube, eight-foot garden, *but* the plants growing under four tubes grow *larger* and *faster* because each plant is receiving more total light energy (compare Figures 34 and 35 at a given distance).

At any output, an eight-foot tube emits more light for a smaller investment (about 25 percent less) than two four-foot tubes. If you can fit an eight-foot fixture, use it in preference to four-foot fixtures. Likewise, a 12-foot lamp works better than an eight. Hardware and lighting stores may not carry HO and VHO fixtures, gro-tubes, or HID lamps, but often will order them. Used tubes are not recommended. The life of each tube is based on use and number of turn-ons. After a year, they produce roughly 80 percent of their original output. The higher the output, the more rapid and severe the decline, and a VHO may produce only 60 percent of original output after a year's use. Incandescents and floodlights are shorter-lived. Replace them after about 500 and 1,500 hours' use, respectively.

Used fixtures are found in junkyards and used office-equipment stores. They are often discarded when commercial buildings are remodeled or torn down. In areas where there is extensive renovation, such as the SoHo section of New York City, people sometimes run across trash bins full of good fixtures.

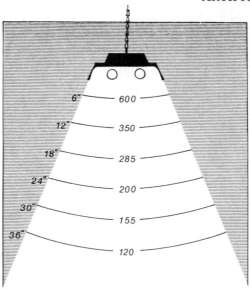

Figure 34. Light output in footcandles from two 40-watt white fluorescents with reflector. *Drawing by E.N. Laincz.*

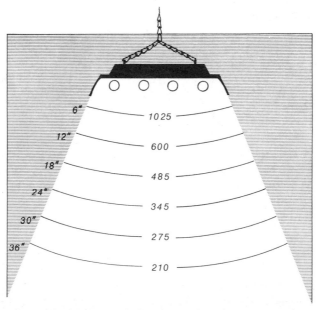

Figure 35. Light output in footcandles from four 40-watt tubes.

Don't use infrared, black lights, ultraviolet (uv), sunlamps or special lamps not designed for human vision or plant growth. These lamps generate most of their energy beyond the visible spectrum and have little effect on photosynthesis. They are a waste of energy and can actually do more harm than good.

There is a myth that great grass comes from the mountains because the air is thin and the plants are exposed to more ultraviolet radiation, which supposedly increases resin production. In well-controlled tests[68] where plants were given an additional 800 watts of light from ultraviolet fluorescents for four hours daily, no significant difference was found in THC content from controls. In fact, the plants under added ultraviolet light actually gave the lowest THC concentrations of plants tested under four different light regimes. Grass is grown in the mountains because the fields are hard to find and the law can't reach them.

Almost any lamp (photo lamps, quartz lamps, etc.) that looks white will help the plants grow. The higher the intensity, the better, as long as concentrated heat doesn't burn the plants. Most white lights benefit by being used with incandescent or floodlights. A filter does not change the light generated. It screens out certain bands to give the color, actually passing less total light energy.

SETTING UP THE GARDEN

Under artificial light, marijuana grows from three to six feet in three months, so the height of the lights must be easy to adjust. Fixtures can be hung from the ceiling, shelves, walls, or from a simple frame constructed for the purpose. If you are hanging the lights from the walls or ceiling, screw hooks directly into a stud. Studs are located in every room corner and are spaced 16, 18 or 24 inches apart. Lights can be supported from lathing using wingbolts, but plaster is too weak to hold a fixture unless a wooden strip held by several wingbolts is attached to the walls or ceiling first to distribute the pressure. Then hang the fixture from a hook in the strip. Closets have hooks and shelves or clothes rungs that are usually sturdy enough to support the fixture. People have gardens under loft beds.

Chains are the easiest means of raising and lowering fixtures. Two chains can be suspended from a solid support from above, and attached to a hook on each end of the fixture. The fixture is then raised by inching the hooks to higher links on the chain. Or rope can be tied to the fixture, through an eye hook or pulley in the ceiling or frame, and knotted to a hook in the wall or frame.

It is also possible to fix the lights permanently and lower the plants on a shelf. The shelf can be suspended from the fixture or raised on a stand. This arrangement is sometimes used in "growing factories" where new plants are started each month, as the older plants are moved to new spaces. This configuration is also well-suited for growing under a skylight. Remember that the plants can be started in small pots and are relatively light the first two months. As the plants grow, the blocks and supports which keep them close to the light are eliminated.

Occasionally it may be convenient to illuminate the plants from the sides. Sometimes fluorescent fixtures are positioned along two sides of the garden. Overhead, floodlights or incandescents are raised as the plants grow.

You might prefer to buy only the end sockets and ballasts and mount the tubes on plywood or a frame of one-by-twos. Mount the sockets so that they are evenly spaced over the garden. This maximizes the efficiency of the lights, since all of the garden space is more equally illuminated (see Figure 38).

Even experienced growers sometimes make the mistake of not using light reflectors. No matter what the light source, its performance in growing the plants will be significantly improved with reflectors. Reflectors direct 25 to 45 percent more light toward the plants. If your fixture is not equipped with a reflector, you can make one from the cardboard boxes in which fixtures and tubes are packaged. Cut off the end flaps and about 10 inches from one end. With staples or tape, face the two wide, inside panels with lengths of aluminum foil. Place the reflector over the fixture and adjust the angle it forms by punching or wedging a pencil between the top edges of the reflecting sides. For VHO or HO fixtures, cut out some holes or flaps in the top of the reflector to allow heat to escape.

Surround the garden with reflective surfaces. These increase the total amount of light the plants receive and also help keep the lower branches illuminated and actively growing when the lights are high. A flat white paint (super white or decorator white) is a better reflecting surface than glossy white paint or

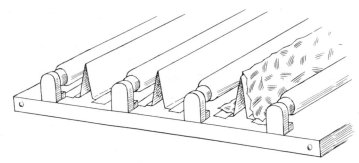

Figure 36. Reflectors can be made from sturdy paper faced with aluminum foil. Make them with staples, tape, or tacks. *Drawings by O. Williams.*

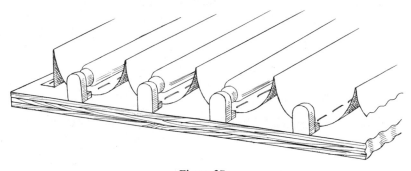

Figure 37.

aluminum foil. Flat white has about three percent more reflecting capacity than aluminum foil, and reflects light more uniformly. The difference is slight, so use whatever means is most convenient. Paint walls that border the garden a flat white or cover them with aluminum foil, mylar, or white posterboard.

Natural-light gardens also benefit from reflectors. Make them out of cardboard painted white or faced with aluminum foil. Once the plants are past the seedling stage, surround them with reflectors; otherwise only one side of the plants will be fully illuminated.

Covering the floor with a plastic dropcloth (about $1 at any hardware store) will protect your floor and your neighbor's ceiling from possible water damage.

Marijuana grows well in a dry atmosphere, but heated or air-conditioned homes are sometimes too dry during germination and early growth. Enclosing the garden in reflectors will contain some of the moisture and insure a healthy humidity.

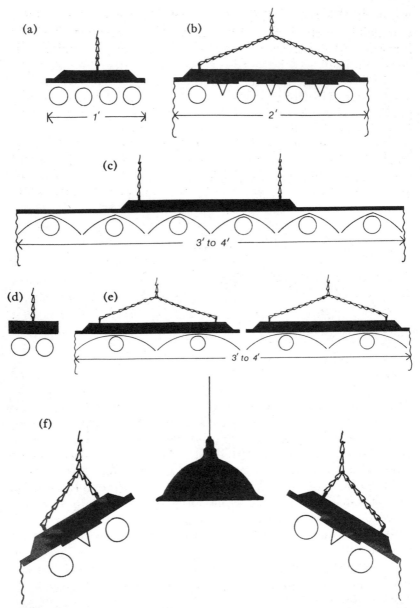

Figure 38. (a) Most standard wattage fixtures come like this. (b) Fixtures with tubes spaced apart are better. (c) Highly recommended: Space six to eight tubes on a piece of plywood. (d) Most VHO fixtures come like this. (e) Better is to mount the tubes separately. (f) Fluorescents can be used to illuminate the plants from the sides with floodlights mounted above. *Drawings by E.N. Laincz.*

White sheet plastic is available to enclose open gardens. Do not completely enclose the garden. Leave some open spaces at the bottom, top and ends of the garden to allow air to circulate. Air circulation will become more important as the plants grow larger. Don't rely on training your pets to stay out of the garden. The garden will attract them, and they can easily destroy young plants by chewing on leaves and stems. Soil is more natural to their instincts than the sidewalk or kitty litter. Protect the garden from pets and toddlers; surround it with white plastic or chicken wire. Large plants are more sturdy and animals can do them little harm. The jungle ambience and an occasional leaf are irresistible to most cats, and they'll spend hours in the garden.

ELECTRICITY

For most growers, the amount of electricity used is of little concern. A four-tube, regular-output, eight-foot fixture draws about 320 watts per hour or about the same as a color TV. The cost increase to your electric bill will be about two to six dollars a month, depending on local rates.

Farmers who devote entire basements or attics to their gardens are sometimes restricted by the amount of current they can draw. Older homes or apartments may have only one 15-ampere circuit but more often have two, for 30 amperes total. Newer homes have either 60 or 100 amperes available through four to six circuits. One 15-ampere circuit can safely accommodate three, two-tube VHO fixtures or six tubes for 1,290 watts, or 16 regular-output, eight-foot tubes for about 1,280 watts total. This allows for a 20 percent safety margin of circuit capacity, which is necessary considering heat loss, starting voltages, etc.

In kitchen and basements the circuits may be rated higher, at either 20 or 30 amperes. You can find out the amperage of the circuit by looking at the fuse rating on the face of the fuse. Determine what room or rooms each circuit is feeding by removing the fuse and seeing which outlets are not working. The wattage capacity of any circuit is found by multiplying volts times amps. Standard United States voltage is 110 to 120 volts.

Fluorescent light fixtures are sometimes sold unwired or

without a line cord, and the job is left to you. Wiring the fixture is very simple and anyone with no previous knowledge of electricity can do it. There is a wiring diagram on the ballast with the wires (four or less), labeled by their color. All you need to do is attach the wires to the lamp sockets as shown in the diagram. Some new fixtures have a socket type with small holes. The bare end of the wire is pushed into the hole and becomes automatically fixed in place. Older-type sockets have the conventional screw terminals.

Indoor gardens may have aluminum foil, chains, and wet or damp floors, all of which are good electrical conductors. Coupled with hanging lights, these conditions could lead to electrical shock. Eliminate any chance of shock whenever you raise or lower the light system by *turning off* the light system. Whenever you change tubes or work on the light system, first pull the plug.

Some fixtures eliminate the risk of dangerous shocks by having a grounding wire (three-pronged plug). If your fixtures are not equipped with a three-pronged plug, you can ground the fixture with a piece of lamp wire. Connect one end to any bare screw on the fixture housing and the other end to the screw that holds the cover plate on an outlet.

Chapter Six
Soil and Containers for It

POTS AND OTHER CONTAINERS

In its natural state, marijuana may grow an extensive root system—a fibrous network of fine, lateral roots that branch off a main, carrot-shaped tap root. In dry areas, the tap root can grow more than six feet deep in its search for water. In moist areas with fertile soil (such as in potting mixtures), the lateral roots are able to supply water and nutritive needs and the tap root remains small, often only three or four inches long on a seven-foot-tall mature plant.

The purpose of the growing medium is to provide adequate water and nutrients in addition to anchoring the roots, which hold the plant upright. By watering and fertilizing as needed, you could grow a six-foot plant in a four-inch* pot or in a three-foot layer of soil over your whole garden; but neither of these extreme procedures is very practical.

Most growers use containers that will hold between two and five gallons of soil. These are a good compromise in terms of weight, space, cost, and labor. They can be moved easily and hold an adequate reservoir of water and nutrients to support a large mature plant.

Some growers use a single large box or several long troughs that hold a six- to 12-inch layer of soil. These have the advantage of minimal restriction of roots and less frequent waterings, but they require more soil and make rotating or moving the plants impractical.

Determine the right size pot to use in your garden by the amount of light per square foot. For a moderately lighted garden (15 to 25 watts per square foot and most window gardens), use one- to three-gallon containers. For gardens with more light energy—over 25 watts per square foot or one-half day or more of sunlight—use three- to eight-gallon containers. The smallest pot we recommend for a full-grown plant is eight inches or one

---

*Pots are measured by diameter across the top.

gallon. This is also a good size for starting plants to be transplanted after two months.

Practically any container that can withstand repeated waterings and has a top at least as wide as its base will do. Each pot must have several holes in the bottom to assure drainage. Growers use flower pots, institutional-sized cans and plastic buckets, baskets and small trash cans, milk crates and wooden boxes.

Plastic trash bags are sometimes used when other large containers can't be found. They must be handled carefully, since shifting the soil damages the fragile lateral roots. They are also more difficult to work with when transplanting. However, a roll of trash bags is an available and inexpensive substitute for other large containers. Plastic bags should be double or triple bagged. Small holes should be punched in the bottom to drain excess water. Use masking tape to patch any unwanted tears. The capacity of the bag should be no more than twice as many gallons as the amount of soil used. For example, with four gallons of soil, the bag should be of a five-gallon, but not more than eight-gallon size. Otherwise, it will not form a cylinder, and the bag will remain a shapeless mass.

Use as many pots as can fit in the lighted area to make the most efficient use of space. Many growers prefer to start the plants in smaller pots, transplanting into larger pots when the plants are larger. There are definite advantages to this method in terms of the yield of the garden, given its space and light energy. Seedlings and small plants take up much less space than they will at maturity, so they can be placed closer together. As the plants grow and begin to crowd each other, remove the less vigorous (to smoke, of course) and transplant the rest into larger pots. Start plants which will be transplanted later in four- to eight-inch flower pots, or one-quart to one-gallon tin cans or milk containers. Peat pots or planting pots are made of compressed plant fiber for the purpose of starting young plants. They are available at garden shops and come in several sizes. Use at least a four-inch pot so that the roots are not restricted in early growth. Peat pots are supposed to break down in the soil, but marijuana's delicate lateral roots may not be able to penetrate unless you score or break away the sides while transplanting. Wax paper cups (six to eight ounces), filled with a soil mixture, work as well as peat pots and are cheaper.

BOX C

Finding Large Containers

Use your ingenuity in finding large containers. Large clay flower pots do not work any better than the large metal and plastic containers discarded by restaurants and food stores. Various milk containers are good starting pots. Many garden shops sell used pots for a few cents each. Wholesalers sell plastic pots by the carton at a discount. Large plastic pots and pails can sometimes be picked up inexpensively at flea markets or variety stores. Any vessel that holds an adequate amount of soil and does not disintegrate from repeated waterings is a satisfactory container.

PROPERTIES OF SOIL

The soil or growing medium serves as a source and reservoir for water, air, and nutrients, and to anchor the roots. Since marijuana grows extremely fast, it has higher water and nutritive needs than most plants grown indoors. The success of your garden depends on supplying the plant with a medium that meets its needs without creating toxic conditions in the process.

There is no such thing as the perfect soil for *Cannabis*. Each variety can grow within a range of soil conditions. For healthy, full growth, marijuana prefers a medium with good drainage, high in available nutrients, and near a neutral pH (7.0). These conditions result from a complex set of physical, chemical and biological factors. We will refer to them simply as: (1) texture; (2) nutrients; (3) pH.

Most indoor growers prepare the growing medium using commercial potting mixes. These mixes are usually sterilized or pasteurized and have good general soil properties. Since they seldom list the contents, nutrients, or pH, do some simple tests of your basic soil whether you buy or dig for it. Then you can adjust the soil to meet the basic requirements of the plant.

Texture

The texture of the medium determines its water-holding and draining properties. Marijuana must have a well-drained medium for healthy growth. Soils that hold too much water or hold it unevenly can drown the roots, leading to poor growth or death of the plant. In a well-drained soil the roots are in contact with air as well as water. Soils that have too much clay, or are overly rich in compost or other organic matter, tend to hold too much water and not enough air. This condition worsens in time. This is especially true of the soil in pots.

You can determine the texture of your soil from its appearance and feel. Dry soil should never cake or form crusts. Dry or slightly moist soil that feels light-weight, airy, or spongy when squeezed, and has a lot of fibrous material, will hold a lot of water. Mix it with materials which decrease its water-holding capacity, such as sand, perlite, or even kitty litter.

Wet soil should remain spongy or loose and never sticky. A wetted ball of soil should crumble or separate easily when poked.

Soil that feels heavy and looks dense with fine particulate matter, or is sandy or gritty, will benefit by being loosened and lightened with fibrous materials such as vermiculite, Jiffy Mix, or sometimes sphagnum moss.

Soil Conditioners to Improve Texture

Perlite (expanded sand or volcanic glass) is a practically weightless horticultural substitute for sand. Sand and perlite contribute no nutrients of their own and are near neutral in *p*H. They hold water, air, and nutrients from the medium on their irregular surfaces and are particularly good at aerating the soil.

Vermiculite (a micaceous material) and *sphagnum moss* contribute small amounts of their own nutrients and are near neutral in *p*H. They hold water, air, and nutrients in their fiber and improve the texture of sandy or fast-draining soils. Jiffy Mix, Ortho Mix, or similar mixes are made of ground vermiculite and sphagnum moss, and are fortified with a small amount of all the necessary nutrients. They are available at neutral *p*H, are good soil conditioners, and are also useful for germinating seeds.

Sphagnum and *Peat Moss* (certain fibrous plant matter) are sometimes used by growers to improve water holding and texture. Both work well in small amounts (10 to 15 percent of soil mixture). In excess, they tend to make the medium too acidic after a few months of watering. Use vermiculite or Jiffy Mix in preference to sphagnum or peat moss.

Nutrients

Nutrients are essential minerals necessary for plant growth. The major nutrients are nitrogen (N), phosphorus (P), and potassium (K), which correspond to the three numbers, in that order, that appear on fertilizer and manure packages, and that give the percentage of each nutrient in the mix (see Chapter 9). Marijuana prefers a medium that is high in nitrogen, and mid-range in phosphorus and potassium. Generally, the darker the soil, the more available nutrients it contains. Commercial soils usually contain a good balance of all nutrients and will support healthy growth for a month or two, even in smaller (one gallon) containers. Many growers prefer to enrich their soil by adding sterilized manures, composts, or humus. All of these provide a good balance of the three major nutrients. They also retain water in their fiber. In excess they cause drainage problems, make the medium too acidic, and attract insects and other pests. A good mixture is one part compost or manure to five to eight parts of soil medium. In large pots (four or five gallons), these mixtures might provide all the nutrients the plant will ever need.

Table 13.

Water Content and Concentration of Nutrients in Manures

Type of Manure	Range	Percentage by Weight of			Availability of Nutrients
		N	P_2O_5	K_2O	
Poultry	Low	1.5	1.0	0.5	Rapid
	High	6.0	4.0	3.0	Rapid
Cow	Low	0.25	0.15	0.25	Medium
	High	1.5	1.0	1.5	Medium

The many prepared organic and chemical fertilizers that can be mixed with the soil vary considerably in available nutrients and concentrations. Used in small amounts, they do not appreciably effect the soil texture. Many prepared fertilizers are deficient in one or more of the major nutrients (see Table 14). Mix them together so there is some of each nutrient, or use them with manures which are complete (contain some of all three major nutrients). When adding fertilizers, remember that organic materials break down at different rates. It is better to use combinations which complement each other, such as poultry manure and cow manure, than to use either fertilizer alone. (See Table 22 in Chapter 13 for a complete list of organic fertilizers.)

Table 14.

Prepared Organic Fertilizers

Type of Fertilizer	Percentage by Weight of			Availability to Plant
	N	P_2O_5	K_2O	
Blood meal	13.0	0	0	Rapid/medium
Bone meal	.5	15.0	0	Medium/slow
Blood/bone meal	6.0	7.0	0	Medium/slow
Cottonseed meal	6.0	2.0	1	Slow/medium
Fish meal	8.0	2.0	0	Slow/medium
Hoof and bone meal	10.0	2.0	0	Slow
Rock phosphate	6.0	24.0	0	Slow
Wood ash	0	1.5	3.0–7.0	Rapid
Greensand	0	0	2.0–8.0	Medium/slow

Chemical fertilizers are made in about every conceivable combination and concentration. Pick one that is complete and where the first number (N) is at least equal to if not higher than both P and K. For example, rose foods may be 12-12-12 or 20-20-20, and work very well for marijuana. Others are: Vigoro 18-4-5 and Ortho 12-6-6. The higher the number, the more concentrated the mix is and, consequently, the more nutrients are available.

Don't use fertilizers which come in pellets or capsules, or that are labeled "timed" or "slow release." They do not work as well indoors as do standard organic and chemical fertilizers. Chemical fertilizers seldom list the amount to mix per pot. You can get some idea by the instructions for application per square foot. Use that amount for each one-half cubic foot of soil mixture.

Many growers add no nutrients at this time but rely on watering with soluble fertilizers when they water. These fertilizers and their application are discussed in Chapter 9.

pH

The pH is a convenient measure of the acidity or alkalinity of the soil medium. It is another way of expressing whether the soil is bitter (alkaline) or sour (acid). The pH is measured on a scale of 0 to 14, with 7.0 assigned neutral; below 7.0 is acid and above is alkaline. You can think of the pH as a measure of the overall chemical charge of the medium. It affects whether nutrients dissolve to forms available to the plant or to forms the plant can't absorb, remaining locked in the soil medium.

Marijuana responds best to a neutral (7.0 pH) medium, although in a fertile, well-drained soil, it will grow well in a range of 6.0 to 8.5. The simplest way to check the pH is with a soil-test kit from a garden shop or nursery. Test kits are chemicals or treated papers—for example, litmus papers or Nitrazine tape—that change color when mixed with a wet soil sample. The color is then matched to a color chart listing the corresponding pH. Nitrazine tape is available, inexpensively, in drug stores. Some meters measure pH, but these are expensive. Agricultural agents, agricultural schools, and local offices of Cooperative Extension will test a soil sample for pH and nutrient content. Occasionally, a garden-shop person will check pH for you or will know the pH of the soils they sell.

Highly alkaline soils are characteristically poor soils that form cakes, crusts, and hardpan. Soil manufacturers don't use them, nor should they be dug for indoor gardens. Alkaline soils are treated with sulphur compounds (e.g., iron sulphate) to lower the pH.

We have never seen commercial soils that were too alkaline for healthy growth, but they are sometimes too acidic. The pH of acid soil is raised by adding lime (calcium-containing) compounds. Liming compounds come in many forms and grades. Some are hydrated lime, limestone, marl, or oyster shells, graded by their particle size or fineness. Use the finest grade available, since it will have more of a neutralizing potential than a coarse grade. You need to use less and are more interested in immediate results than long-term soil improvement. For indoor gardens, use hydrated lime (available in any hardware store) or wood ashes to raise the pH. Hydrated lime is rated over 90 percent for its neutralizing potential. Wood ashes will neutralize soil acids roughly one-half as well as hydrated lime. However, they also contain some nutrients (potassium, phosphorus, magnesium, and micronutrients) and are handy and free.

There is no exact formula we can give you for raising the pH. The pH does not have to be exact; it's an approximation. At low pH it takes less lime to raise the pH one point than it does when the pH is near neutral. Sandy soils need less lime to raise the pH one point than soils high in clay or organic matter. In general, add three cups of hydrated lime or six cups of fine woodash to every large bag (50 pounds or a cubic foot) of soil to raise the pH one point. For soils that test slightly acid (about 6.5), add two cups of lime or four cups of wood ash.

Soil that tested below 6.0 should be retested in about two weeks, after thoroughly mixing and wetting the soil. Repeat the application until the pH is in an acceptable range. Check the pH of plain water to see if it is influencing the tests. Distilled water is neutral, but tap water sometimes has minerals that can change the pH. Hard water is alkaline. Sulphurous water and highly chlorinated water are acidic.

If you have already added lime to a soil that now tests from 6.5 to 7.0, don't add more lime trying to reach exactly 7.0. Too much lime will interfere with nutrient uptake, notably of potassium, phosphorus, and magnesium.

General Soil Characteristics

The texture, pH, and available nutrients of the soil are all related. The most important single factor is texture (good drainage). When soil drains poorly, it creates anaerobic (without air) pockets in the soil. Bacteria or microbes that live without air will begin to multiply and displace beneficial microbes that need air to survive. The anaerobic microbes break down organic matter to a finer consistency, and release CO_2 and organic acids to the medium. Drainage worsens, the acids lower the pH, and nutrients, even though present, become unavailable to the plant.

The result can be a four-month-old marijuana plant that is only three inches tall, especially if you use high concentrations of manures and composts, peat and sphagnum moss. If your soil lists manures or composts as additives, add no more than 10 percent of these on your own.

Drainage problems sometimes develop after several months of healthy growth. It is a good idea to add about 20 percent sand or perlite to even a well-drained soil. You can never add

Figure 39. A four-month-old plant grown in toxic soil. (Chronological age has little to do with the plant's development.)

too much of these; they can only improve drainage. They dilute the nutritive value of the soil, but you can always water with soluble fertilizers.

Mixtures using many components in combination seem to work particularly well. This may be because, at a micro-level each component presents a slightly different set of physical, chemical and biological factors. What the plant can't take up at one point may be readily available at another.

PREPARING COMMERCIAL SOILS AND MIXES

Garden soils (or loams) and potting mixes are actually two different groups of products, although they are frequently mislabeled. Some companies sell soil in large bags and a potting mixture in smaller bags, while labeling them the same. Soils and potting mixtures are usually manufactured locally, since transportation costs are prohibitive; so they differ in each area.

Texture and Nutrients

Soils and loams are usually topsoil blended with humus or compost for use as a top dressing in gardens, for planting large outdoor containers, or for the soil part of a potting mixture. They may have a tendency to compact under indoor conditions and will benefit from the addition of perlite or vermiculite. Soils and loams usually contain a good supply of nutrients and may support a full-grown plant in a large container. Commercial soils that are heavy generally work better than lightweight soils. Heavy soils usually contain topsoil, in which marijuana grows very well. Lightness indicates more fibrous content.

For examples of possible soil mixtures, see Box D.

BOX D

Examples of Soil Mixtures*

1. 5 parts soil
   2 parts perlite
   1 part cow manure

2. 8 parts soil
   3 parts sand
   ¼ part 10-10-10 chemical fertilizer

3. 5 parts soil
   2 parts perlite
   2 parts humus
   ½ part cottonseed meal

4. 4 parts soil
   1 part sand
   1 part vermiculite
   2 parts humus
   ½ part poultry manure

5. 3 parts soil
   1 part perlite
   1 part sand
   2 parts Jiffy Mix
   ½ part blood/bone meal
   ½ part wood ash

6. 6 parts soil
   2 parts perlite
   2 parts vermiculite
   ½ part poultry manure
   ½ part cow manure
   1 part wood ash

*Almost all fertilizers are acidic, and need to be neutralized by lime. For the above mixtures, or any similar ones, mix in one cup of lime for each five pounds of manure, cottonseed meal, or chemical fertilizer in order to adjust the pH.

Potting mixes are intended to support an average-size house plant in a relatively small pot. They are sometimes manufactured entirely from wood and bark fiber, composts, and soil conditioners. These mixes are made to hold a lot of water and slowly release nutrients over a period of time, which is what most house plants require. For marijuana, these mixes seldom contain enough nutrients to support healthy growth for more than a couple of months. (Their N is usually low, P adequate, and K usually very high.) They work best when sand or perlite is added to improve drainage, and fertilizers are added to offset their low nutrient content.

The pH

Most commercial mixes and soils are between 6.0 and 7.0 in pH, a healthy range for marijuana. If you buy your soil, it will not be too alkaline for healthy growth, but it might be too acidic. You can minimize the chances of getting an acid soil by avoiding soils with "peat" or "sphagnum" in their names. Avoid soils that are prescribed for acid-loving plants such as African violets or azaleas, or for use in terrariums. With common sense, you can buy a soil, add two cups of lime to each large bag, and not have to worry about the pH. However, the surest procedure is to test the pH yourself.

Probably the best way to find the right soil for your garden is to ask long-term growers. They can relate their past experiences with various mixes and blends. Most long-term growers with whom we have talked have tried many of the mixes available in their areas. A reliable, enlightened nurseryperson or plant-shop operator may also be able to give you some advice.

BUYING SOIL COMPONENTS

All the materials discussed here are available at farm and garden stores or nurseries. Many suburban supermarkets sell large bags of soil and humus. Always buy your materials in the largest units possible to reduce the cost.

Large bags of soil and humus come in either 50-pound bags or one- to four-cubic-foot bags. A 50-pound bag fills about six gallons. There are eight gallons to a cubic foot. Perlite is sold in four-cubic-foot bags (thirty-two gallons). Jiffy Mix and vermiculite are sold in four-cubic-foot bags and in 16-pound bags (about 18 gallons). Sand, perlite and vermiculite come in coarse, medium, and fine grades. All grades work well, but if you have a choice, choose coarse. Sand (not beach sand) is an excellent soil conditioner. The only disadvantage is its heavy weight. Buy sand from lumber yards or hardware stores where it is sold for cement work. It will cost from 1/50 to one-half the cost of garden or horticultural sand. Sand from piles at construction sites works very well.

Calculating the Amount of Soil

The maximum amount of soil mixture for any garden can be found by multiplying the capacity of the largest pot you plan to use by the number of pots that you can fit in the garden. In many cases, the actual amount of the mixture used will be somewhat less. Two illustrations follow.

   1. A small garden with a two-tube, eight-foot fixture (160W). Using 20 watts per square foot for fast growth gives 160W divided by 20W/sq. ft. = eight sq. ft. The *largest* pot needed for this system is three gallons, but two gallons would work. You can fit about 10 three-gallon pots in eight square feet; so 3 X 10 = 30 gallons of soil mixture are needed (see Box E).

BOX E
Examples Showing How Much Soil Material to Buy to Fill a Known Number of Unit-Volume Containers

Example 1. For a garden eight square feet in size,

Buy	Component	Which amounts to
3 50-lb (6 gal. ea.) bags of	soil	18 gallons
1 cubic foot of	perlite	8 gallons
30 lbs of	humus	3 gallons
10 lbs of	chicken manure	2 gallons
	Total	31 gallons

Example 2. For a garden 24 square feet in size,

Buy	Component	Which amounts to
4 1-cu. ft. bags of	soil	32 gallons
2 1-cu. ft. bags of	perlite	16 gallons
1 1-cu. ft. bag of	vermiculite	8 gallons
20 pounds of	cow manure	3 gallons
	cottonseed meal	2 gallons
	wood ash	2 gallons
	Total	63 gallons

2. A large garden with two two-tube, eight-foot VHO fixtures (four times 215 watts or 860 total watts) illuminating a garden three by eight feet, or 24 square feet.

860 watts divided by 24 sq. ft. = about 36 W/sq. ft.

The largest pot size for this system is about five gallons. About 16 five-gallon containers can fit in 24 square feet; so 16 X 5 = 80 gal. of mixture are needed. But you could start many more plants in smaller containers and transplant when they are root-bound. You do not use more soil by starting in smaller pots, since all soil is reused. In many cases, you actually use much less soil.

In this system you could start and fit about 40 plants in one-gallon pots in 24 square feet. When the plants begin to crowd each other, some are harvested, making room for the others, which are transplanted to larger pots. In practice, a high-energy system such as this one (36W/sq. ft.) will grow large plants whose size is limited mainly by the space available. Twelve large female plants are about the most you would want in the system during flowering and for final harvest. Sixty gallons of mixture is all that is needed for the seedlings and the mature crop. This is one-fourth less than the original estimate of 80 gallons, and you actually will harvest a lot more grass (see Box E).

Mixing and Potting

Mix your soil in a large basin, barrel, or bathtub. Individual pots are filled with mixtures by using a smaller container to measure out by part or volume.

Perlite, sand, and dry soil can give off clouds of dust. When mixing large amounts of these, wear a breathing mask or handkerchief over your nose and mouth.

To pot any of the mixtures, first cover any large drainage holes with a square of window screen or newspaper to prevent the mixture from running out. Place a layer of sand, perlite, or gravel about one inch deep to insure drainage. Fill the pots with soil mixture to within three-fourths of an inch from the top of the pot. If your mixture contains manures or composts, cover the last inch or two in each pot with the mixture minus the manure and compost. This will prevent flies, gnats, molds, and other pests from being attracted to the garden. Press spongy soils firmly (not tightly) to allow for more soil in each pot;

otherwise, after a period of watering, the soil will settle and the pot will no longer be full.

Some growers add a few brads or nails to each pot to supply the plant with iron, one of the necessary nutrients. Water the pots and allow them to stand for a day or two before planting. As the soil becomes evenly moist, beneficial bacteria begin to grow and nutrients start to dissolve.

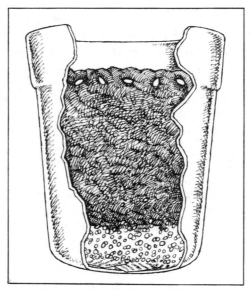

Figure 40. *Drawing by O. Williams.*

DIGGING SOIL

Most growers prefer to buy their soil, while some prefer to dig it. Marijuana cannot tolerate heavy clays, mucks, or soils that dry to crusts. Choose a soil from a healthy garden or field, or from an area that supports a lush growth of annual weeds.

Fields that support a good crop of alfalfa, corn or other grains will support a good crop of marijuana. Fields with beets, carrots, and sugar cane indicate a well-drained soil, with near neutral pH. Red clover, sweet clover, and bluegrass have soil requirements similar to those of marijuana. Garden soils are usually fertile and well-drained, but often need lime to counteract soil acidity.

Take the topsoil layer that starts about two inches below the surface debris. Good soil will look dark, feel moist, and smell clean and earthy. Use all of the topsoil layer that maintains its dark color and is interlaced with roots. Your hands should be able to easily penetrate the underlying topsoil if the soil is in good condition. When the soil changes color, or roots no longer are apparent, then you are past the fertile topsoil layer. Abundant worms, millipedes, and other small lifeforms are a good indication that the soil is healthy. A rich layer of topsoil collects by walls, fences, and hedges where leaves and debris collect and decay to a rich humus. Sift the soil to remove stones and root clods. Also, shake out the root clods, which are rich in nutrients.

Soil that is dug should be tested the same way as already prescribed. It should be adjusted with at least 30 percent sand or perlite (vermiculite for very sandy soils), since potting will affect the drainage of even well-drained soils. Never use manures or composts that are not completely degraded to a clean-smelling humus.

Soil that is dug must be sterilized to kill weed seeds, insect eggs, and harmful molds and fungi. Some chemical treatments (e.g., formaldehyde) are mixed with water and poured over the soil to sterilize it. Soil can be sterilized in a pressure cooker at 15 pounds pressure for 15 minutes, or by baking wet soil in a large pot at 200 degrees for 30 to 40 minutes. Be advised that baking soil will release some formidable odors.

GROWING METHODS

As we said before, there are probably as many growing methods as there are marijuana growers. These methods are personal preferences or adaptions to fit particular situations; one method is not necessarily better than any other. However, the value of a garden is often based on the amount of high-quality grass it yields. Since indoor gardens are limited in size, you want the plants to quickly fill the garden with lush growth in order to use the garden efficiently. Otherwise, for the first couple of months the lights are shining on empty space.

Secondly, the possession of small quantities of marijuana will probably be decriminalized nationally within the next few years. Decriminalization for personal possession will open the way for decriminalization for cultivation for personal possession. But small quantities are more difficult to define for cultivation than for simple possession, which is done by weight. Several possible ways to limit the amount for cultivation have been raised: by the number of plants, by the area cultivated, or by the number of plants at a particular stage of development. The outcome may determine whether you try to grow the largest plants possible or the most plants possible in a given area.

There are several ways to increase your garden's yield.

1. Pinch or cut back the growing shoots when the plants are young. This forces each plant to develop several strong growing shoots and generally yields large robust plants.

2. Plant a number of plants in each pot.

3. Start many plants in small pots and transplant the best plants to larger pots when the plants crowd each other.

4. Use different light systems to grow plants at different growth stages.

Here are some examples of how to carry out each of these four methods.

1. Fill the growing area with large containers (about five gallons each). Start several plants in each pot but thin the seedlings over a period of six weeks to two months, until one plant is left in each pot. During the fourth or fifth week of growth, pinch back the plants to about equal heights. Cut the growing

Figure 41. Plant clipped at fourth internode.

shoot at about the fourth internode. Each plant will develop a sturdy stem which will support four to eight growing stems and will quickly fill any empty space in the garden. The whole garden is then treated like a hedge. After another month or two, you cut back the growing shoots again to have plants of equal heights. Remove the male plants as soon as they begin to release pollen (or before any male flowers open for sinsemilla). This will leave more space and light for the females to develop. By the time females flower, they've been cut back two or three times or more, and form a dense growth of growing shoots that fill the garden with a cubic layer of flowers. Some growers maintain the plants for up to a year before the final harvest.

2. This method also requires large pots. Instead of thinning the seedlings to leave one per pot, leave at least three. After a few months of growth, remove any plants that lag far behind or any plants that show male flowers. The value of this method is that the odds are at least seven to one that any pot will have at least one female plant.

Most of the plants you'll grow will fill out with branches by four months at the latest. Often the branches develop on young seedlings. The plants may begin to look like small Christmas trees by the second to third months of growth.

Generally, you don't want to have more than three or four

Figure 42. Basement growing factory in Atlanta.

plants in a five-gallon container, because growth will be limited by competition for light and space.

Some varieties never do fill out. The branches remain small, only two to three inches long, and yield very little grass. We've seen plants like this grown from grass from Vietnam, Thailand, Afghanistan, and Africa. These plants are also quite short, being four to six feet tall fully grown. With varieties like this, it is better not to pinch tops, and to start about six plants per square foot of garden space. At harvest, the garden will be crowded with top stems that are laden with flower clusters.

Of course, you don't know what varieties will look like until you've seen them grow. For *most* varieties, each plant will need *at least* one square foot of space at maturity. It is much less common to find varieties that naturally grow small or especially thin, and, therefore, are those of which you would want to plant more than a few per large pot.

3. Another popular way to grow is to start plants in a large number of small pots. As the plants crowd each other, some are removed and the rest transplanted to larger pots.

4. To get the most for your investment requires conservation of light and soil. When the plants are young, a large number fit into a small place. Some growers take advantage of this fact by having several light systems, each with plants at different growth stages. The plants are rotated into larger gardens and pots. This method conserves space, materials, and electricity, and yields a harvest every two months. Using this method, "growing factories" turn out a steady supply of potent grass.

Table 15.

Age and Size of Garden

Months of Growth	No. of Plants	Size of Containers	Size of Garden	No. of Tubes[a] in Lighting System
0 - 2	24	1 gallon	12" X 8'	2
2 - 4	24	3 gallon	30" X 8'	4
4 - 6	12[b]	5 gallon	36" X 8'	6

[a]Each tube is 8 feet long and is rated at 80 watts.
[b]During months 4 to 6, the flowering stage, male plants can be removed; they will, on the average, constitute half the initial population.

Chapter Seven

Maintaining the Correct Environment

REQUIREMENTS FOR GERMINATION

Before the seed fell, almost all of its water was sapped to prepare the seed for winter. With only the tiny drop that it holds, the embryo lives a life so slow as to be outside of time as we know it. *Cannabis* seeds need only water to germinate or sprout. The seeds germinate without light and at temperatures low enough to form ice. Higher temperatures hasten germination. Fresh, homegrown Oaxacan seeds germinated in three days at 70°F and in eight days at 33°F. Temperatures 70° to 90°F are best for germination.

Fresh, mature seeds have a high rate of germination (about 90 to 100 percent) and sprout quickly. Usually sprouts appear three to seven days after planting. Older seeds (over a year, depending on storage) have a lower rate of germination and respond slower. They may take up to three weeks to sprout. To get an idea of what to expect from the seeds, follow the procedure on p. 47.

Seeds that do sprout will grow normally, no matter how old they are or how long they take to sprout. From any batch of seeds, most of the ones that sprout will do so within two or three days of each other. A few will continue to come up as many as six months later, but the garden should consist of plants that are basically the same age and size. This makes the garden easier to care for.

Choosing Seeds

Different varieties grow at different rates and attain different sizes and shapes. Under artificial lights, gardens planted from one batch of grass require the least attention, because the plants sprout and grow uniformly and can all be tended at the same time. When several varieties are grown together, some plants are taller than others; you must adjust the height of the plants to

Figure 43. Within each seed lies an embryo.

keep the marijuana equally illuminated. You may also have to water and fertilize the plants on an individual basis. Some growers start different varieties under separate light systems. On the other hand, planting several varieties offers you a comparison in potency and yield, and a source for hybrids if you want to develop seed. The next time you plant you'll know which seeds gave the best results and what growing methods will work best for you.

There is no strict correlation between the form and height of the plants and seed size, color, or pattern. However, some large-seeded varieties grow too tall, with long spaces between leaves. Under artificial lights they yield more stems than leaves. If you have a choice between two equally potent grasses, and one has particularly large seeds (3/16 to 1/4 inch), choose the smaller-seeded variety.

Sowing

The easiest way to start the plants is to sow the seeds directly into the soil. First, wet the soil with a moderate amount of water, enough to get the soil evenly moist without water running out the bottom. This takes about one-half quart of water for one-gallon containers, and about one quart for three-gallon containers. Plant the seeds a quarter- to half-inch deep. The germination rate is lower when they are planted deeper; and if seeds are planted less than one-quarter inch deep, the sprouts may have difficulty anchoring their roots. Plant about six seeds per pot to assure some sprouts in each pot. Gently press each seed into the soil. Cover the seeds with soil and sprinkle lightly with water. Each day, sprinkle or spray the surface with enough water to thoroughly wet the top half-inch of soil, since the seed must be kept moist for germination.

For most people, germinating the seed is easy. Problems with germination come from either too much or too little water. If you saturate the pots with water, and especially if you continue to saturate the pots after the seeds have sprouted, the seedlings may develop stem rot or root problems. When stem rot develops, the base turns brown, and the seedlings fall over, ending the garden. This can also happen if you keep the seedlings in germination boxes or terrariums where the humidity is very high. When the humidity is low, the soil surface dries out quickly and the seeds won't germinate. Sprouts that may come up shrivel and dry at the base of the stem and die.

The key to germination is to keep the soil surface moist after first having moistened the whole pot; then, after the first sprouts have been up for a few days, let the surface of the soil dry between waterings. Don't spray the surface any more. Water with medium amounts of water when the soil in the top couple of inches feels dry. For small pots, water seedlings about twice a week. For larger pots, once each week or two may be enough.

Some growers prefer to plant only seeds they know will sprout, especially when planting seeds which have a low viability. Start the seeds in wet towels or a glass of water. Add one teaspoon of liquid bleach (a three-percent solution) to each cup of water. This will prevent fungus from attacking the seeds, which happens when they are soaked for more than three days. Check the seeds each day. Plant when the radical or root begins to come out from the pointed end of the seed. *Cannabis* seed is quite small and has only enough stored food for the embryo to anchor its root and raise its cotyledons. The more developed the root is when planted, the less energy it has to anchor itself in the soil. The sprout may die or growth be delayed until the root is established (transplant shock). In Figure 44, the seeds in a

Figure 44. Seeds in a circle are ready to plant. Center sprout will not survive planting.

circle are all ready to plant. The center seed will not survive transplanting.

Some growers prefer to start the plants in a germination box. This extra hassle is not necessary. Transplanting seedlings from one medium to another often causes transplant shock. It is best to plant the seeds directly into the soil.

If you use soilless mixtures, your seedlings should be started in paper cups, peat pots, or other small pots filled with a soil mixture (see "Transplanting" in Chapter 8). This procedure is also helpful if you have difficulty starting the plants in large containers. Expandable peat pellets also work very well.

The position of the seed in the soil has a slight effect on germination. The root directs its growth in response to gravity, as shown in Figure 45. However, germination is a little faster when the seeds are planted with the pointed end up. The difference is small, and it's not really necessary to position the seeds in the soil.

If a dry atmosphere presents problems, you can create the moist atmosphere of a germination box and still plant directly in the pot. Cover the seeds with transparent plastic cups or glasses, or cover the pot with plastic kitchen wrap. This creates

Figure 45. The root directs its growth toward gravity. Seeds are germinated between glass and cotton, and held vertically. Four seeds to left have pointed end up. Two middle seeds are horizontal. Six seeds on right have pointed end down.

a greenhouse effect and keeps the soil surface moist without watering. Remove all the covers *as soon* as you see the first sprouts begin to appear; the sprouts will die if the cover is left on.

Figure 46. During germination soil can be kept moist by using plastic covers to create a greenhouse effect.

LIGHT CYCLE AND DISTANCE OF LIGHTS FROM PLANTS

The seed doesn't need light to germinate. The sprout does need light as soon as it breaks through the soil. Most growers turn the lights on when they sow the seeds, though, to warm the soil and encourage germination. Lights may also dry the surface of the soil, especially in large pots or with VHO fixtures. If this is a problem during germination, leave the lights off until you see the first sprout breaking through the soil; or hang the lights about 18 inches above the soil, and lower them to six inches as soon as the sprouts appear.

It is important for normal development that the plants receive a regulated day/night cycle. We emphatically recommend that you use an automatic electric timer (about $8). A timer makes gardening much easier, since you don't have to turn the lights on or off each day. The plants won't suffer from irregular hours or your weekend vacations. Set the timer so that the plants get about 16 to 18 hours of light a day, and leave it on this setting until the plants are well grown (three to six months) and you decide to trigger flowering.

During the seedling and vegetative stages of growth, the plants may be subjected to light during their night period. During flowering, however, the night period must be completely dark.

The plants grow more slowly with less than 16 hours of artificial light a day, and they may flower prematurely. Some growers leave the lights on up to 24 hours. A cycle longer than 18 hours may increase the growth rate, especially if the plants are not saturated with light. A longer cycle is helpful in small gardens, such as under standard four-foot fixtures.

No matter what the light source, place the lights as close to the tops as possible without burning the plants. Pay no attention to the manufacturer's instructions for the distance of the plants from the lights; these instructions don't apply to a high-energy plant such as *Cannabis*. With standard-wattage tubes, keep the lights from two to six inches above the plant tops. With VHO tubes, allow four to eight inches. Maintain the lights at these distances throughout the life of the garden. In most cases you will have to raise the lights once or twice a week as the plants grow.

Standard fluorescents don't get hot enough to burn the plants unless they are in direct contact with leaves for several hours. VHO tubes will burn leaves before they touch them. But you do want to keep the lights as close to the plants as possible. This encourages stocky, robust growth. Incandescents and floodlights get very hot; place them at a greater distance from the plants. Test the distance by feeling for heat with your hands. Place the bulb at the distance where you begin to feel its heat. For a 75-watt incandescent lamp, this is about eight inches.

WATER

Water, the fluid of life, makes up more than 80 percent of the weight of the living plant. Within the cells, life processes take place in a water solution. Water also dissolves nutrients in the soil, and this solution is absorbed by the roots. About 99 percent of the water absorbed passes from the roots into the conduits (xylem) of the stem, where it is distributed to the leaves via the xylem of the leaf veins. Transpiration is the evaporation of water from the leaves. The flow of water from the soil, through the plant to the air, is called the transpiration stream. Less than one percent of the water absorbed is broken down to provide electrons (usually in the form of hydrogen) which, along with carbon dioxide, are used to form carbohydrates during photosynthesis. The rest of the water is transpired to the air.

Watering

Water provides hydrogen for plant growth, and also carries nutrients throughout the plant in the transpiration stream. However, it is not true that the more water given a plant, the faster it will grow. Certainly, if a plant is consistently underwatered, its growth rate slows. However, lack of water does not limit photosynthesis until the soil in the pot is dry and the plant is wilting.

The amount of water, and how often to water, varies with the size of the plants and pots, soil composition, and the temperature, humidity, and circulation of the air, to name a few variables. But watering is pretty much a matter of common sense.

During germination, keep the soil surface moist. But once the seedlings are established, let the top layer of soil dry out before watering again. This will eliminate any chance of stem rot. Water around the stems rather than on them. Seedlings are likely to fall over if watered roughly; use a hand sprinkler.

In general, when the soil about two inches deep feels dry, water so that the soil is evenly moist but not so much that water runs out the drainage holes and carries away the soil's nutrients. After a few trials, you will know approximately how much water the pots can hold. Marijuana cannot tolerate a soggy or saturated soil. Plants grown in constantly wet soil are slower-growing, usually less potent, and prone to attack from stem rot.

Overwatering is a common problem; it develops from consistently watering too often. When the plants are small, they transpire much less water. Seedlings in large pots need to be watered much less often than when the plants are large or are in small pots. A large pot that was saturated during germination may hold enough water for the first three weeks of growth. On the other hand, a six-foot plant in a six-inch pot may have to be watered every day. Always water enough to moisten all the soil. Don't just wet the surface layer.

Underwatering is less of a problem, since it is easily recognized. When the soil becomes too dry, the plant wilts. Plant cells are kept rigid by the pressure of their cell contents, which are mostly water. With the water gone, they collapse. First the bottom leaves droop, and the condition quickly works its way up the plant until the top lops over. If this happens, water immediately. Recovery is so fast, you can follow the movement of water up the stem as it fills and brings turgor to the leaves. A plant may survive a wilted condition of several days, but at the very least some leaves will drop.

Don't keep the pots constantly wet, and don't wait until the plant wilts. Let the soil go through a wet and dry cycle, which will aerate the soil and aid nutrient uptake. Most growers find that they need to water about once or twice a week.

When some soils get particularly dry, the water is not absorbed and runs down the sides and out the bottom of the pot. This may be a problem the first time you water the soil, or if you allow the soil to get very dry. To remedy, add a couple of drops of liquid detergent to a gallon of water. Detergent acts as a wetting agent and the water is absorbed more readily. First water each pot with about one cup of the solution. Allow the pots to stand for 15 minutes, then finish watering with the usual amount of pure water.

Use tepid water; it soaks into the soil more easily and will not shock the roots. Try to water during the plant's morning hours. Water from the top of the pot. If you do want to water from the bottom with trays (not recommended), place a layer of pebbles or gravel in the trays to insure drainage. Don't leave the pots sitting in water until the pot is heavily saturated. The water displaces the soil's oxygen, and the plants grow poorly.

Tap water in some areas is highly chlorinated, which does not seem to harm *Cannabis;* and many fine crops are raised with water straight from the tap. But chlorine could possibly affect the plants indirectly, by killing some beneficial microorganisms in the soil. Chlorine also makes the water slightly acidic. However, neither effect is likely to be serious. Some growers have asked whether they should use pet-shop preparations that are sold to remove chlorine from water in fish tanks. These preparations generally add sodium, which removes the chlorine by forming sodium chloride (table salt). This solution does not harm the plants, although repeated use may make the soil too saline. Probably the best procedure is to simply allow the water to sit in an open container for a few days. The chlorine is introduced to water as the gas Cl_2, which dissipates to the air. The water temperature also reaches a comfortable level for the plants.

Hard (alkaline) water contains a number of minerals (e.g., Ca^{++}, Mg^{++}, K^+) which are essentially nutrients to the plants. Water softeners remove these minerals by replacing them with sodium, which forms slightly salty water. It is much better to water with hard water, because artificially softened water may prove harmful after some time. Occasionally, water may be acidic (sulphurous). Counteract this by mixing one teaspoon of hydrated lime per quart of water and watering with the solution once a month.

Water and Potency

We've seen no studies that have evaluated potency in relation to water. A few studies have mentioned the fact that plants that received less water were slightly more potent. Water stress has been practiced by several marijuana-growing cultures. In parts of India, watering is kept to a minimum during flowering.

To limit watering, water with the usual amounts but as infrequently as possible. To encourage good growth, yet keep watering to a minimum, wait until the plants are a few months old before you curtail watering. Give the plants their normal water and note the number of days before they begin to wilt; then regularly water on the day before you expect them to wilt. As the plants get larger, the water needs increase, but this generally stabilizes by the time of flowering.

AIR

The properties of the air seldom present any problems for indoor gardeners. The plants grow well under the ordinary conditions that are found in most homes and can withstand extremes that are rarely found indoors. The plants can survive, in fact thrive, in an atmosphere many house plants can't tolerate. For plant growth, the most important properties of the air are temperature, humidity, and composition.

Temperature and Growth Rate

Temperature control should be no problem. The plants can withstand temperatures from freezing to over 100°F. Plant growth is closely related to temperature. Marijuana varieties are, in general, adapted to warm if not hot climates. Different varieties will reach their maximum rate of photosynthesis at different temperatures. For almost all marijuana varieties, the rate of photosynthesis will increase sharply with increases in temperatures up to about 70°F. Some strains reach their peak rate of growth at about 75°F. Others, especially from areas near the equator, such as Colombia, may not reach their peak rate until the temperature is about 90°F. However, for all varieties, increases in the growth rate will be slight with increased temperatures over 75°F. The average temperature for maximum growth is about 75° to 80°F. In other words, normal household temperatures are fine for growing marijuana and no special temperature control is necessary for most gardens.

Don't set up the garden right next to, or in contact with, a heat source such as radiator or furnace. If the garden is *nearby,* the plants should do quite well. The plants are most susceptible to cool temperatures during germination and the first few weeks of growth. In basement gardens, the floor temperature is often lower than the air. It is a good idea to raise the pots off the floor with pallets or boxes. The seeds will germinate quicker, and the plants will get off to a faster start.

If heating is necessary, propane catalytic heaters work well, are safe and clean, and increase the carbon-dioxide content of the air. Electric and natural gas heaters also work well. Do not use kerosene or gasoline heaters. They do not burn cleanly, and the pollutants they produce may harm the plants. Any heater that burns a fuel must be clean and in good working order. Otherwise, it may release carbon monoxide, which is more dangerous to you than to the plants.

Temperature and Potency

Since marijuana varieties are most often grown in semi-tropical and tropical areas, the idea that high temperatures are necessary for potent marijuana is firmly entrenched in marijuana lore. This myth, like many others, is slowly disappearing as marijuana farmers and researchers accumulate more experience and knowledge. There are only a few published papers on the effects of temperature on potency. The best study we've seen[19] grew four different varieties in a controlled environment under artificial lights on a 15-hour daylength. Two temperature regimes were used: a "warm" regime, with temperatures of about 73°F during the day and 61°F at night (about average for most homes); and a "hot" regime, set at 90°F daytime and 73°F at night. In all four varieties, the concentration of THC and of total cannabinoids was higher under the "warm" regime. For instance, a Nepalese strain was 3.4 times higher in concentration of total cannabinoids, and 4.4 times higher in THC, when grown under the "warm" regime than the same strain grown under the "hot" regime. Although we agree with the findings in principle, these figures are higher than our own experience tells us.

Interpretation of the data does show one point clearly. In all four varieties, the amount of THC lost as CBN was higher under the "hot" regime (see Table 16), even though the *concentration* of THC was higher under the "warm" regime.

Another research group in France has looked at the relationship of potency to temperature. The most recent paper[79] compared four temperature regimes, given in descending order of potencies found: 75°F day, 75°F night (highest potency); 72°F day, 54°F night; 81°F day, 81°F night; and 90°F day, 54°F

Table 16.

**Percentage of THC Oxidized to CBN in Plants Grown
Under "Hot" and "Warm" Temperature Regimes**

	Place of Origin of Variety			
Regime	Panama	Jamaica	Nepal	Illinois
"Hot" (90° F – 73° F)	14.3	21.4	27.7	9.8
"Warm" (73° F – 61° F)	6.5	2.2	10.8	0

night (lowest concentration of THC). In each, the day period was 16 hours and the night period eight hours.

Interestingly, this same research group in an earlier paper[20] reported that the concentration of THC was higher for male plants grown at 90°–72°F than for those grown at 72°–54°F. For the female plants, the differences in THC concentration were small. The variety used was a propyl variety (type IV) containing about half as much THCV as THC. For both the male and female plants, the concentrations of THCV were higher under the 90°–72°F regime.

The simplest interpretation of all these results is that mild temperatures seem to be optimum for potency. Temperatures over 90°F or below 60°F seem to decrease the concentration of THC and total cannabinoids. Also, at higher temperatures, much more THC will be lost as CBN. And last, propyl varieties may produce less THCV under a cool regime. Bear in mind that none of these papers accounted for all of the many variables that could have affected the findings. For instance, the concentration of THC was 18 times higher at 75°–75°F than at 90°–54°F. We've never seen differences of this magnitude, and sampling errors undoubtedly influenced the findings.

In terms of growth rate and potency, daily temperatures of about 75°F, give or take a few degrees, are roughly optimum. Normal household temperatures are in the low 70's during daytime and the low 60's at night. The heat from a light system will raise the garden's temperature a few degrees. In most gardens temperatures will be near 75°F during the day. Nighttime temperatures generally drop about 10 to 15 degrees. When nighttime temperatures drop into the 50's or below, set the light cycle to turn on during the early morning, when the temperature will be lowest. In a small room, the light system will generate enough heat to warm the garden without any need for a heater. Whenever you wish to raise the temperature by, say, five or 10 degrees, it is better to add more lights than a heater. The plants will benefit from the additional light, as well as from the heat they generate. And an electric heater, watt for watt, doesn't generate much more heat than a lamp and its fixture.

Composition of the Air

Air provides two essential ingredients for the living plant: oxygen and carbon dioxide. The plant uses oxygen for respiration in the same way we do. The oxygen is used to burn carbohydrates (CH_2O) and other food, yielding energy (ATP; see Chapter 4) for the organism, and releasing carbon dioxide and water into the environment.

During photosynthesis, CO_2 is used to form carbohydrates. As part of photosynthesis, light energy is used to split water molecules, releasing oxygen into the environment. In plants, the net result from respiration and photosynthesis is that much more oxygen is released than consumed, and more carbon dioxide is consumed than released. The oxygen in the Earth's atmosphere is formed by photosynthetic organisms.

The similarity between plant and animal respiration *ends* at a cellular level. Plants don't have lungs to move the air. The passage of gases, whether oxygen or carbon dioxide, is primarily a passive process. The gases diffuse through microscopic pores called stomata, found in *Cannabis* on the undersides of the leaves. The plants can open and close their stomata, allowing moderate control of the flow of air. However, for good exchange of gases, the plants require adequate ventilation for air circulation.

Cannabis is not particularly susceptible to a stuffy or stagnant atmosphere. A garden in the corner of a room that is open to the house will be adequately ventilated. Ventilation is not a problem unless the garden is large and fills a quarter or more of the space in a room. Gardens in small, confined spaces such as closets must be opened daily, preferably for the duration of the light period. Plants growing in a closed closet may do quite well for the first month, but they'll need the door opened as the plants begin to fill the space. The larger the plants get, the greater the need for freely circulating air.

When the weather is mild, an open, but screened, window is the best solution for ventilation. In large indoor gardens where there isn't much air circulation, a small fan is helpful. After germination, make spaces in the surrounding reflectors to allow air to circulate freely. Leave spaces at the bottom, ends, and the top of the garden. The higher the temperature or the humidity, the more the plants need good ventilation.

CO_2

The concentration of carbon dioxide in the atmosphere is generally about 0.03 percent; in cities it is a third higher, about 0.04 percent. Plants can use more carbon dioxide than is supplied by the ordinary atmosphere. Increases to about 0.5 percent concentration increase the rate of growth as long as other factors such as low light levels are not limiting. You can increase the carbon-dioxide content of the air by keeping the plants in a closed room at night. The plants' respiration during the night increases the concentration of carbon dioxide. Simply visiting with your plants increases their carbon-dioxide supply.

In large commercial greenhouses, carbon dioxide sometimes is added to increase growth rates. Carbon dioxide is slowly released from tanks at controlled rates, disperses over the plant tops and, being heavier than air, settles slowly downward. High rates of carbon dioxide can be toxic to you. We do not recommend additional carbon dioxide unless you have a large installation where you can carefully control the supply. Additional carbon dioxide will increase the growth rate but probably will not increase potency.

Smoke and Pollution

Cannabis does not seem to be susceptible to what has become everyday pollution. The plants naturally colonize roadsides and city dumps. The common air pollutant, carbon monoxide, is believed to cause a higher incidence of female plants in a population. Statistical studies of roadside stands of *Cannabis* (where higher carbon-monoxide concentrations are due to car exhaust) showed that populations averaged about 55 to 60 percent females instead of the normal 50 percent. Carbon monoxide is deadly. There are more efficient ways of increasing the number of females, and we strongly advise you not to try to use carbon monoxide on the plants.

Smoke is not healthy to living things, whether it comes from tobacco, marijuana, a wood fire, or a charcoal-broiled steak. *Cannabis* survives in smokey homes quite well. However, in confined areas with very heavy concentrations of smoke, say during a party, the plants may actually die. Male plants seem the more susceptible. You should provide a fresh air source for your plants, if you must smoke in the growing area. And move the plants to another room if you're having a party.

HUMIDITY

Marijuana flourishes through a wide range of relative humidity. It can grow in an atmosphere as dry as a desert or as moist as a jungle. Under ordinary household conditions, the humidity will rarely be too extreme for healthy growth. The effects of the humidity on plant growth are closely tied to temperature, wind speed, and the moisture of the soil.

The relative humidity affects the rate of the plant's transpiration. With high humidity, water evaporates from the leaves more slowly; transpiration slows, and growth then slows also. With low humidity, water evaporates rapidly; the plant may not be able to absorb water fast enough to maintain an equilibrium, and will protect itself from dehydration by closing its stomata. This slows the transpiration rate and growth also slows. There is a noticeable slowing of growth because of humidity only when the humidity stays at an extreme (less than 20 percent or over 90 percent).

Cannabis seems to respond best through a range of 40 to 80 percent relative humidity. You should protect the plants from the direct outflow of a heater or air conditioner, both of which give off very dry air. During the first few weeks of growth, the plants are especially susceptible to a dry atmosphere. If this is a problem, loosely enclose the garden with aluminum foil, white sheet plastic, or other materials. This will trap some of the transpired moisture and raise the humidity in the garden. Once the seedlings are growing well, the drier household atmosphere is preferred.

Where the humidity is consistently over 80 percent, the plants may develop stem rot or grow more slowly. Good air circulation from open windows or a small fan is the best solution.

As long as the air is freely circulating, the plants will grow well at higher humidities. Dehumidifiers are expensive (over $100) and an extravagance.

Humidity and Potency

As far as we know, there has been little work done correlating the relative humidity with potency. In the two related cases we've seen,[85,117] neither study was intended to examine the effects of relative humidity and potency. However, a lower humidity (50 to 70 percent) produced slightly more potent plants than a higher relative humidity (80 percent and over).

A dry atmosphere seems to produce more potent plants. When the humidity is about 50 percent or less, plant development is more compact, and the leaves have thinner blades. When the atmosphere is humid, growth is taller and the leaves luxuriant with wider blades. The advantage to the plant is that wider blades have more surface area and hence can transpire more water. The converse is that thinner blades help conserve water. Higher potency may simply be due to less leaf tissue for a given amount of cannabinoids and resin glands.

The temperature also influences the form and size of the leaves. At higher temperatures, the leaves grow closer together; under a cool regime, the leaves are larger, have wider blades, and are spaced farther apart.[77] Possibly, cool temperatures yield slightly lower potency for much the same reason that a moist atmosphere does.

However, differences in potency caused by any of the growth factors (light, nutrients, water, temperature, humidity, etc.) are small compared to differences caused by the variety (heredity) and full maturation (expression of heredity). For example, the humidity in Jamaica, Colombia, Thailand, and many other countries associated with fine marijuana is relatively high and averages about 80 percent.

However, try to keep the atmosphere dry. The atmosphere in heated or air-conditioned homes is already dry (usually 15 to 40 percent). For this reason, many growers sow so that the plants mature during the winter if the home is heated or in midsummer if it is air-conditioned. As we mentioned, there should be no need to use dehumidifiers. Good air circulation and raising the temperature to 75° to 80°F are the simplest means of dealing with high humidity.

Chapter Eight
Gardening Techniques

THINNING

Depending on the viability of the seeds, there should be several plants growing in each pot. Most growers thin to one plant per pot, but the plants don't have to be thinned until they crowd each other and have filled the garden with foliage. The longer you let them grow, the more potent they'll be.

It is virtually impossible to tell the gender of the plants when they are young. The normal ratio of males to females in *Cannabis* is one to one. Some farmers end up with more male plants because of their thinning practices. When the plants are less than a month old, the male plants often appear taller and better developed than the females. The male seedling uses more of its energy to develop its aboveground parts than the female. The female devotes more energy to establishing a strong root system. During the first few weeks, don't thin the plants by leaving only the tallest, or you'll wind up with a higher ratio of males. Try to leave seedlings that are healthy and vigorous and that are roughly at the same point of development.

To thin your garden, remove any plants with yellow, white, or distorted leaves. Remove the less vigorous and those that lag far behind in development. Cut the unwanted plants near the base; the root system can remain in the pot.

These harvested seedlings will be your first taste of home-grown grass. Usually they produce a mild buzz, but if you separate the growing tips from the large leaves, they may be more potent.

TRANSPLANTING

However you transplant, try to disturb and expose the roots as little as possible. If you transplant carefully, the plants will not exhibit delayed or slowed growth due to transplant shock.

Transplanting Seedlings

When the plants are a week to two weeks old, transplant to any pot that has no plants. First, moisten the soil in the pot from which you will remove the transplant and let it sit for a few minutes. Take a spade or a large spoon, and insert it between the transplant and the plant that will be left to grow. Try to leave at least one inch of space from spoon to stem. Lever the spoon toward the side of the pot, in order to take up a good-size wedge of soil. Place the transplant in a prepared hole at the same depth that it was growing before. Replace the soil in both pots and moisten lightly again to bond the new soil with the original. If you are careful, a wedge of soil can be removed intact. The root system will not be disturbed and the plant will survive with little or no transplant shock. Do not fertilize a transplant for two weeks.

To prevent possible drop-off and wilting from shock, you may want to use Rootone or Transplantone. These safe powders, available at nurseries, contain root-growth hormones and fungicides. They won't be necessary if you transplant carefully.

Transplanting to Large Pots

Transplanting from smaller to larger pots is a simple procedure. The marijuana root system quickly fills small pots. To transplant, moisten the soil and let it sit to become evenly moist. Pick the potted plant up, and, while holding the base of the stem, rap the pot sharply against something solid. You might cover the soil surface with a piece of newspaper or aluminum foil, which makes the job cleaner. When it is done at the right time, the root system, with all the soil adhering, will pop out of the pot intact.

An approximate time guide for transplanting is shown in Table 17. At these times, give or take a week, the plants should be rootbound and all the soil will adhere to the roots, making the transplanting clean and easy.

If the root system has not filled the pots by this time, wait a few weeks and the process will be easier. If the root system comes out in a small ball and much of the soil is empty of roots, then soil conditions are poor (usually poor drainage and overwatering) or you are transplanting much too soon.

If the root system doesn't easily pop out, run a knife around the sides of the pot. Sometimes the roots stick to the sides, particularly in paper and clay containers. Check to see if the drainage holes are plugged. Plugged holes stop air from displacing the soil, and the vacuum pressure prevents the soil from sliding out of the pot.

Table 17.

Guide for Transplanting

Transplant	During
Six-ounce cups	Second to third week
Four-inch pots	Third to fourth week
Six-inch pots (half gallon)	Fourth to fifth week
Eight-inch pots (one gallon)	Seventh week to eighth week
Two-gallon containers	About the tenth week

Transplant into a soil mixture that is the same as (or is very like) the one in the original pots. Otherwise, the soils may have different osmotic properties, and the water may not disperse evenly. (This doesn't apply to small pots that are used for germination and are filled with vermiculite, Jiffy Mix, or other mediums.) Don't bury the stem. Keep the stem base at the same depth that it was growing.

Figure 47. Transplant when the plant is rootbound.

Transplanting in Plastic Bags

To transplant plants that are in plastic bags, place the old bag into the larger-size bag. Put some soil mixture underneath, to bring the base of the stem to where the new soil surface will be. Cut the old plastic bag away and fill the side spaces with soil mixture. Two people make the job easier.

SUPPORTS FOR PLANTS

Under natural conditions, stems undergo stress from wind, rain, and animals. These stresses, which indoor plants do not ordinarily face, strengthen the stem. Indoor stems grow sturdy enough to support their own weight and not much more. Plant energy is used to produce more light-gathering leaf tissue, rather than wind-resisting stem tissue. Stems remain slender, usually about one-half to three-quarter inches at maturity. Since you are growing the plants for their leaves and flowers, this does not present a problem.

Healthy plants do not ordinarily need support. If many of your plants have weak or spindly stems, there is a deficiency in either light or nutrients (notably potassium). Simply not having enough light will cause the plants to elongate, with sparse foliage and weak growth. Too much red light will cause elongation, too, so make sure you include a strong blue light, if you are using incandescents or floodlights.

Hanging the lights higher than the recommended distances will cause the plants to elongate by rapidly growing up to the lights. Unlike sunlight, the intensity of artificial light diminishes dramatically with the distance from the lights. The plants respond by growing toward the light, seeking the higher intensity.

Under artificial light, some plants may need support during the seedling stage or because of accident. Depending on plant size, use straws, pencils, dowels, or standard plant stakes such as cane sticks. Set them in the soil and affix the stem with string, masking tape, or wire twists such as those that come with plastic trash bags. Do not tie string or wire tightly around the stem; make a loose loop. The stem will grow in girth and can be injured by a tight loop.

Probably the simplest method of support is to take a rigid piece of wire, form a "C" at one end and bend it to a right angle to the stem. Set the straight end in the soil and place the stem inside the "C." Pipe cleaners are ideal for seedlings. With larger plants, straighten a coat hanger and use the same method.

A common practice in greenhouses where tree seedlings are raised is to shake each plant once or twice daily. This practice simulates natural vibrations from the wind, and the plant reacts by increasing the growth around the stem. The stem grows thicker and stronger, and the tree can better fend once it is transplanted. It works the same way with marijuana. A fan blowing on the plants will also work. These practices are useful if you plan to move your plants outdoors. Otherwise, healthy indoor plants that will remain indoors need no special stem strengthening.

UNIFORM GROWTH

The light intensity from artificial light drops dramatically as the distance from the light source increases. When the plants are not of equal height, the shorter ones receive less light and consequently grow slower than the taller ones. This compounds the situation and, left to themselves, the shorter plants will stop growing and eventually die from lack of light.

It is important to keep all of the plants close to the lights.

Figure 48. Hang the fixture at an angle corresponding to that of the tops of the plants. *Drawing by O. Williams.*

This encourages stocky, full growth and can make the difference between harvesting stems and harvesting smoking material.

One way to deal with uneven height is to line the plants up by height and hang the light system at an angle corresponding to the line of the plant tops. As the plants grow, move them to different spots in the garden to accommodate their different sizes. Or raise the shorter plants up to the lights by placing them on milk crates, tin cans, bricks, etc.

The quality and quantity of light emitted by a fluorescent is strongest in the middle and weaker toward the ends of the tube. Female plants require more light than males. Once the genders of the plants become clear, move the males to the ends of the system, thus leaving the stronger middle light for the females.

PRUNING

Probably the easiest way to deal with uneven growth is to cut back the taller plants to the average height. You may find this emotionally difficult, but pruning will not harm the plant. Cutting off the growing shoot forces the plant to develop its branches. Some growers cut back all of their plants when they are three to four weeks old. Any horizontal space is quickly filled with growing branches and the plants grow full and robust.

The growing shoots are the most potent plant parts until the flowers appear. Generally, the potency increases with growth. By three months' age, most shoots will be high-quality smoke. You can cut shoots at any time; just don't overdo it. Give the plant a chance to grow and fill out to a good size. Severe pruning will slow growth. New growth may be distorted and abnormal, with a drop in potency.

Each time you cut a growing shoot, whether it is the stem tip or a branch tip, two shoots begin to grow from the nearest leaf axils. However, don't think that cutting all the growing shoots of a plant twenty times over the course of a season will yield a plant bearing over a million new shoots, or even that the plants will double their size if pruned. Pruning simply allows the plant to develop its branches earlier. The branches present more area to gather light and, hence, can grow to fill a larger space. However, the plant's size is basically determined by the seed's potential within the limitations of the environment.

Cutting the growing shoots or removing some leaves does not harm the plants. Plants are well adapted to the loss of parts to predators, wind, etc., in the natural world. When leaves are damaged or lost, the plant plugs the wound. The leaf isn't replaced or repaired, but new leaves are continually being formed from the growing shoots. The stem, since it connects all parts of the plant, is more important to the plant as a whole. When the stem breaks or creases, it is capable of repair. You can help the plant repair its stem by splinting the wound or somehow propping the stem up straight. Stems take about four or five days to heal.

When you cut the stem or leaves, you may see the plant's sap momentarily spurt before the wound is plugged. The sap contains primarily the products of photosynthesis, in the form of sucrose (table sugar). Smaller amounts of materials associated with the living organism such as minerals, amino acids, and enzymes are also present. In marijuana, the sap is usually colorless, although a bright red color—it looks like blood—is not uncommon in later life. The red color is due to hematin compounds and anthocyanin pigments that naturally build up in some varieties. The red color may also indicate a nutrient deficiency, notably of nitrogen, phosphorus, potassium, or magnesium.

TRAINING

Plants grow from the tips of their stems and branches. The growing tip (apical meristem) of the plant contains a hormone that acts as a growth inhibitor. This prevents the branches (lateral buds) from growing. The further a branch is from the growing tip, the less the effect of the inhibitor. This is why some species of plants form a cone or Christmas-tree shape with the longest branches toward the bottom of the stem. This is also why the branches grow from the top of the plant when the tip is removed. Once the growing tip is removed, the next highest growing shoot(s) becomes the source of the inhibitor. Under artificial light, the bottom branches may not receive enough light to grow even though they are far away from the inhibitor. Usually the longest branches are toward the middle of the plant.

Some growers hate to cut the growing shoot on the main stem, since it forms the largest and most potent buds by harvest. But you can neutralize the effects of the inhibitor, without cutting the growing shoot, by bending the tip. This allows you to control the height of the plants, and forces them to branch. The top two to six inches of the stem are flexible. Bend it in an arc and secure it to the stem with a wire twist or string. Remove the wire twist in a few days so that the growing tip does not break itself as it twists up to the light. Don't bend the stem too far down. Keep it in the strong light or else it will stop growing. If you accidently break the tip, you can splint it with matchsticks or ice-cream sticks secured with wire twists or tape until it heals.

Figure 49. The flexible tip is held in place with a wire twist.

To develop large, full plants with well-developed branches, secure the growing tip once or twice for a few days while the plants are young (one to three months).

It is possible to train the tip so that the stem will form a series of "S" shapes or even circles. During flowering, train the tips so that they grow horizontally. This method encourages thick, dense growth. The branch tips can also be trained. Keep bending any tips that grow above the others. This creates a garden filled with a cubic layer of *vigorous* flower clusters rather than a lot of stems.

We want to emphasize that when you get the knack of training the tips, you can more than double the yield of the most potent plant parts.

Figure 50. Stem trained in an "S" shape.

Figure 51. Tops trained horizontally during flowering.

Chapter Nine
Nutrients and Fertilizing

NUTRIENTS

There are about 15 elements known to be essential to plant life. Carbon, hydrogen, and oxygen are absorbed from air and water. The remaining 12 elements are absorbed primarily from the soil, in mineral (inorganic) forms such as NO_3^- and K^+. They constitute a natural part of soil that becomes available to the plant as organic matter decays and soil particles such as sand and clay dissolve.

Soil elements that are necessary for normal growth are called nutrients. The elements nitrogen (N), phosphorus (P), and potassium (K) are considered major nutrients. The three numbers that appear on all fertilizer packages give the available percentage of these three nutrients that the fertilizer contains; and always in the order N-P-K. For example, 10-2-0 means 10 percent N, 2 percent P (actually, 2 percent P_2O_5), and no K (actually, no K_2O). Fertility is often measured by the amounts of major nutrients a soil contains. Relatively large amounts of N-P-K are needed for lush growth.

Three other elements—calcium (Ca), sulfur (S), and magnesium (Mg)—are called secondary nutrients. Plants require less of these nutrients, and most cultivable soils contain adequate amounts for good growth.

Six remaining elements are called trace elements or micronutrients. As their name implies, they are needed in very small amounts. Commercial soils contain enough trace elements to sustain normal growth. The trace elements are also present in manures, humus, ash, and limestone.

Nitrogen

The amount of nitrogen a soil can supply is the best single indication of its fertility. Nitrogen, more than any other soil nutrient, is inextricably linked with the living ecosystem. Nitrogen is continually cycled through living systems: from soil to plants and back to the soil, primarily by the activity of soil micro-organisms. Nitrogen is essential to all life.

Nitrogen is a key element in the structure of amino acids, the molecules which make up proteins. These, and all other biomolecules, are synthesized by the plant. Chlorophyll, genetic material (for example, DNA), and numerous enzymes and plant hormones contain nitrogen. Hence, N is necessary for many of the plant's life processes.

Cannabis is a nitrophile, a lover of nitrogen. Given ample N, Cannabis will outgrow practically any plant. Ample nitrogen is associated with fast, lush growth, and the plant requires a steady supply of nitrogen throughout its life. Marijuana's requirements for N are highest during the vegetative growth stages.

Phosphorus

P is a constituent of energy-transfer compounds such as NADP and ATP, and molecular complexes such as the genes. The energy compounds are necessary for photosynthesis, respiration, and synthesis of biomolecules. Cannabis takes up large amounts of P during germination and seedling stages. During flowering and seed set, Cannabis' need for phosphorus is also high.

Potassium

K influences many plant processes, including photosynthesis and respiration, protein synthesis, and the uptake of nutrients. Just as with P, K uptake is highest during the earliest growth stages. K is associated with sturdy stems and resistance to disease in plants.

Calcium

Ca functions as a coenzyme in the synthesis of fatty compounds and cell membranes, and is necessary for normal mitosis (replication of cells). Plants take up much more Ca than the small amount necessary for normal growth. Ca is not added to soil as a nutrient; it is added to adjust the soil's chemistry or pH.

Sulfur

S is a constituent of certain amino acids and proteins. It is an important part of plant vitamins, such as biotin and thiamine, which are necessary for normal respiration and metabolism.

(Plants synthesize all the vitamins they need.) Most soils suitable for growing marijuana contain plenty of S.

Magnesium

Mg is involved in protein synthesis and metabolism of carbohydrates. Mg is the central element in the structure of chlorophyll molecules and hence has an important role in photosynthesis. Most mineral soils and commercial soils have a good supply of Mg.

Trace Elements

The trace elements (Fe, Mn, Mb, B, Cu, Zn) are particularly important in the coenzymes and catalysts of the plant's biochemistry. Many life processes, particularly the synthesis and degradation of molecules, energy transfer, and transport of compounds within the plant, depend on trace elements. Trace elements are not used in large quantities to spur growth, but are necessary in minute amounts for normal growth. Indoor soils rarely require an addition of trace elements.

All the nutrients are needed for normal growth. However, most of them are supplied by the potting soil. Ca, S, and the trace elements rarely present any problems. For most growers, fertilizing will simply require periodic watering with a complete fertilizer, one that contains N, P, and K.

APPLICATION: FERTILIZING

To grow to a large size, marijuana requires a steady supply of nutrients. These can be added to the soil before planting or anytime during growth. Bulk fertilizers are added while the soil is mixed, as described in Chapter 6. These include manures, composts, humus, and concentrated fertilizers, such as rose food. Once the plants are growing, never condition or mulch indoor soils with bulk fertilizers. They promote molds and fungi and attract other pests to the garden. Concentrated fertilizers can damage the plants if they come in direct contact with the stem or roots.

While the plants are growing, nutrients are given in solution; they are dissolved in water, and the plants are watered as usual.

Soluble fertilizers can be either organic or inorganic (chemical), and come in a wide range of concentrations and proportions of nutrients. Two organic fertilizers are liquid manure (about 1.5-1.0-1.5) and fish emulsion* (about 5-1-1). Chemical fertilizers commonly may have 20-20-20 or 5-10-5, or may contain only one nutrient, such as 16-0-0.

A 10-5-5 fertilizer is 20 percent soluble nutrients and 80 percent inert ingredients. A 30-10-10 has 50 percent available nutrients and 50 percent inert ingredients. There is approximately the same amount of N in one tsp. of 30-10-10 as in three tsps. of 10-5-5.

Actually, you can use almost any fertilizer, but the nitrogen content should be proportionately high, and there should be some P and K also present. For example, a 20-20-20 would work fine, as would a 12-6-6 or a 3-4-3, but not a 2-10-10 or a 5-10-0.

How much fertilizer to use and how often to fertilize depend primarily on the fertility of the soil and the size of the container relative to the size of the plant. Small plants in large pots usually do not need to be fertilized. Even in small pots, most plants do not need to be fertilized for at least the first month.

As the plants grow, they take nutrients from the soil, and these must be replaced to maintain vigorous growth. During the vegetative stage, even plants in large pots generally require some fertilizing, particularly with N.

The rate of growth of indoor plants is usually limited by the amount of light and space, once adequate nutrients are supplied. At this point, an increase in nutrients will not increase growth. Your goal is to supply the plants with their nutritional needs without overfertilizing and thus toxifying the soil.

Most fertilizers are designed for home use and have instructions for fertilizing houseplants. Marijuana is not a houseplant, and it requires more nutrients than houseplants. The extra nutrients that it needs may be supplied by the use of large pots and a fertile soil mixture. In many cases, you will need to fertilize only in the dosages recommended on fertilizer packages for houseplants. For instance, Rapid-Gro (23-19-17) is popular among marijuana growers; use one tablespoon per gallon of water every two weeks.

A typical program for fertilizing might be to fertilize during

---

*Some fish emulsion may contain whale by-products.

the fifth week of growth and every two weeks thereafter until flowering. Then discontinue fertilizing (or give at one-half concentration) unless the plants show a definite need for nutrients. It is better to fertilize with a more diluted solution more often than to give concentrated doses at longer intervals. (For instance, if instructions call for one tablespoon of fertilizer per gallon once a month, use one-quarter tablespoon per gallon once a week.)

Make sure that a fertilizer is completely dissolved in the water before you apply it. Put the recommended amount of fertilizer in a clear glass bottle and mix with about one cup of water. Shake vigorously and then allow it to settle. If any particles of fertilizer are not dissolved, shake again before adding the rest of the water. If you have difficulty getting all the fertilizer to dissolve, first add hot tap water. If the fertilizer still does not completely dissolve, you should use another fertilizer.

Never fertilize a dry soil or dry soilless medium. If the medium is dry, first water with about one-half quart of plain water per pot. Let the pots sit for about 15 minutes so that the water is evenly dispersed in the pot. Then fertilize as usual.

It is difficult to give instructions for fertilizing that will cover all garden situations. You want to supply the plant with its nutritive needs, but overfertilizing can toxify the soil. Fertilizing according to instructions for houseplants (both in frequency and concentration) should not toxify the soil. However, the plants may sometimes require more frequent or more concentrated fertilizing. A good way to judge the plants' needs is to not fertilize one plant, double the fertilizer of another plant, and give the rest of the plants their normal dose. If the unfertilized plant grows more slowly, or shows symptoms of deficiencies, then probably all the plants are depending on soluble fertilizers and must be fertilized regularly. If the plant receiving the double dose grows faster than the other plants, increase the other plants' supply also. On the other hand, if there is little difference among the plants, then the soil is providing the plants with enough nutrients, and they either should not be fertilized or should be fertilized with a less-concentrated solution.

Because they are grown in a relatively small area, it is easy to overfertilize indoor plants. When plants are vigorous, look healthy, and are growing steadily, don't be anxious to fertilize, particularly if you have already fertilized several times with soluble fertilizers. Slow growth or symptoms of deficiencies clearly indicate the need for fertilizing.

Overfertilizing

In an effort to do the best for their plants, some people actually do the worst. Overfertilizing puts excessive amounts of nutrients in the soil, causing toxic soil conditions. Excessive amounts of one nutrient can interfere with uptake of another nutrient, or change normal plant-soil relations. Since it takes time for a build-up to occur, high concentrations of nutrients generally encourage excellent growth until the toxic level is reached.

It takes less N than other nutrients to toxify the soil; hence there is less margin for error when using N. Too much N changes the osmotic balance between plant and soil. Instead of water being drawn into the plant, water is drawn away and the plant dehydrates. The leaves feel limp even though the plant is well watered. The plant will soon die. The tips of the leaves die first and very rapidly the leaves change color, usually to gold, but sometimes to brown or green-grey. This change in the plants is faster, more dramatic, and more serious than for any kind of nutrient deficiency.

You can save the plants by immediately leaching the pots as soon as the condition is recognized. Place the pots outdoors or in a sink or bathtub. Discard the top inch or two of loose dirt. Run lukewarm water through the soil until a gallon of water for each two gallons of soil has passed through each pot. The leaves recover turgor in one or two days if the treatment works.

Foliar Feeding

Foliar feeding* (spraying the leaves with fertilizer) is a good way to give the plants nutrients without building up the amount of soluble substances in the soil. After the first month, foliar feed the plants with, for example, fish emulsion or a chemical fertilizer. Use any fertilizer that states it can be used for foliar feeding even if it says "not recommended for foliar feeding

---

*Nitrogen fertilizers are usually NO_3 (nitrate) or NO_2 (nitrite), substances which are also used to preserve food. They have been shown to undergo reactions to form carcinogenic substances (nitrosoamines). As with eating food treated with nitrates and nitrites (hot dogs, sandwich meats, etc.), there is a possibility that such substances might be ingested by eating or smoking foliar-fed plants.

houseplants." Use a fine-mist sprayer, such as a clean Windex or Fantastik bottle. Dilute the fertilizer according to directions (fish emulsion at one tablespoon per gallon) and spray both sides of the leaves. When foliar feeding, you should spray the plants with plain water the next day, to dissolve unabsorbed nutrients and clean the plants.

Foliar spraying is also a good way to treat plants suffering from nutrient deficiencies. Some nutrient deficiencies actually are caused by the soil's chemistry, rather than by absence of the nutrient in the soil. Addition of the necessary nutrient to the soil may not cure the plants' problem, because the nutrient becomes locked in the soil, or its uptake may be limited by high concentrations of other elements present in the soil. Foliar feeding is direct, and if the plant's deficiency symptoms do not begin to clear up, then the diagnosis is probably incorrect.

NUTRIENT DEFICIENCIES

Before Diagnosing

Before you assume the plant has a nutrient deficiency, make sure the problem is not due to other causes. Examine the plant carefully for plant pests, especially on the undersides of damaged leaves, and along the stem and in the soil.

Even under the best conditions, not all leaves form perfectly or remain perfectly green. Small leaves that grew on the young seedling normally die within a month or two. Under artificial lights, bottom leaves may be shielded from the light, or be too far away from the light to carry on chlorosynthesis. These leaves will gradually turn pale or yellow, and may form brown areas as they die. However, healthy large leaves should remain green *at least* three to four feet below the plant tops, even on those plants under small light systems. Under low light, the lower-growing shoots as well as the large leaves on the main stem are affected. Some symptoms of nutrient deficiencies begin first at the bottom of the plant, but these symptoms generally affect the lower leaves on the main stem first, and then progress to the leaves on the branches.

Although some deficiency symptoms start on the lower, older leaves, others start at the growing shoots or at the top of the plants. This difference depends on whether or not the nutrient is mobile and can move from the older leaves to the active growing shoots. Deficiency symptoms of mobile nutrients start at the bottom of the plant. Conversely, deficiency symptoms of immobile nutrients first appear on the younger leaves or growing shoots at the top of the plant. N, P, K, Mg, B, and Mb are mobile in the plant. Mn and Zn are less mobile, and Ca, S, Fe, and Cu are generally immobile.

A dry atmosphere or wet soil may cause the blade tips to turn brown. Brown leaf tips also may indicate a nutrient deficiency, but in this case, more tissue will turn brown than just the end tips.

Chlorosis and necrosis are two terms which describe symptoms of disease in plants. Chlorosis means lacking green (chlorophyll). Chlorotic leaves are pale green to yellow or white. Chlorotic leaves often show some recovery after the necessary nutrient is supplied. Necrosis means that the tissue is dead. Dead tissue can be gold, rust, brown, or grey. It is dry and crumbles when squeezed. Necrotic tissue cannot recover.

Symptoms of deficiencies of either N, P, or K have the following in common: all involve some yellowing and necrosis of the lower leaves, and all are accompanied by red/purple color in stems and petioles. The simplest way to remedy these deficiencies is to fertilize with a complete fertilizer containing nearly equal proportions of the three nutrients.

Nitrogen

N is the most common deficiency of *Cannabis* indoors or out. Nitrogen deficiencies may be quite subtle, particularly outdoors, where the soil may continuously provide a small amount of nitrogen. In this case the top of the plant will appear healthy, and the plant will grow steadily, but at a slow pace. The deficiency becomes more apparent with growth, as more and more of the lower leaves yellow and fall. The first sign is a gradual, *uniform* yellowing of the large, lower leaves. Once the leaf yellows, necrotic tips and areas form as the leaves dry to a gold or rust color. In small pots, the whole plant may appear pale (or lime color) before many bottom leaves are affected to the point that they yellow or die. Symptoms that accompany N deficiency include red stems and petioles, smaller leaves, slow growth, and a smaller, sparse profile. Usually there is a rapid yellowing and loss of the lower leaves that progresses quickly to the top of the plant unless nitrogen is soon added.

Remedy by fertilizing with any soluble N fertilizer or with a complete fertilizer that is high in N. If your diagnosis is correct, some recovery should be visible in three or four days. Pale leaves will regain some color but not increase in size. New growth will be much more vigorous and new stems and petioles will have normal green color.

Indoors, you should expect plants to need N fertilization a few times during growth. Once a plant shows a N deficiency, you should fertilize regularly to maintain healthy and vigorous growth. Fertilize at about one-half the concentration recommended for soilless mixtures. Increase the treatment only if the plants show symptoms again. Once the plants are flowering, you may choose not to fertilize if the plants are vigorous. They will have enough stored N to complete flowering and you don't want to chance toxifying the soil at this late date.

Phosphorus

P deficiency is not common indoors, but may appear outdoors, particularly in dry, alkaline soils or in depleted soils, or during cool weather. Phosphorus deficiency is characterized by slow and sometimes stunted growth. Leaves overall are smaller and dark green; red color appears in petioles and stems. The leaves may also develop red or purple color starting on the veins of the underside of the leaf. Generally the tips of most of the leaf blades on the lower portion of the plant die before the leaves lose color. Lower leaves slowly turn yellow before they die. Remedy with any soluble P-containing fertilizer. Affected leaves do not show much recovery, but the plant should perk up, and the symptoms do not progress.

Potassium

K deficiencies sometimes show on indoor plants even when there is apparently enough supplied for normal growth. Often, potassium-deficient plants are the tallest* and appear to be the most vigorous. Starting on the large lower leaves, the tips of the blades brown and die. Necrotic areas or spots form on the blades, particularly along the margins. Sometimes the leaves are

---

*Potassium is associated with apical dominance in some plant species.

spattered with chlorotic tissue before necrosis develops, and the leaves look pale or yellow. Symptoms may appear on indoor plants grown in a soil rich in organic material. This may be due to high salinity (Na) of some manures or composts used in the soil. Red stems and petioles accompany potassium deficiencies. K deficiencies that could seriously affect your crop rarely occur with indoor soils. However, mild symptoms are quite common. Usually the plants grow very well except for some necrotic spotting or areas on the older leaves. (This condition is primarily an aesthetic problem, and you may choose not to fertilize. See page 281.)

K deficiencies can be treated with any fertilizer that contains potassium. Wood ashes dissolved in water are a handy source of potassium. Recovery is slow. New growth will not have the red color, and leaves will stop spotting after a couple of weeks. In a K-deficient soil, much of the added potassium is absorbed by the soil until a chemical balance is reached. Then additional potassium becomes readily available to the plant.

Calcium

Ca deficiencies are rare and do not occur if you have added any lime compound or wood ash. But calcium is added primarily to regulate soil chemistry and pH. Make sure that you add lime to soil mixtures when adding manures, cottonseed meal, or other acidic bulk fertilizers. An excess of acidic soil additives may create magnesium or iron deficiencies, or very slow, stunted growth. Remedy by adding one teaspoon of dolomitic lime per quart of water until the plants show marked improvement. Periodically fertilize with a complete fertilizer. Foliar feeding is most beneficial until the soil's chemistry reaches a new balance.

Sulfur

S is plentiful in both organic and mineral soils. Liming and good aeration increases S availability. Hence S deficiencies should not occur in soils that are suitable for growing marijuana. However, sulfur deficiencies sometimes can be confused with N deficiencies and may also occur because of an excess of other nutrients in the soil solution. Sulfur-deficiency symptoms usually start at the top of the plant. There is a general yellowing of the new leaves. In pots, the whole plant may lose some green color. Both sulfur and Mg deficiencies can be treated with the same compound, epsom salts ($MgSO_4$). Epsom salts, or bathing salts, are inexpensive and available at drug stores.

Magnesium

Mg deficiencies are fairly common. They frequently occur in soilless mixtures, since many otherwise all-purpose fertilizers do not contain Mg. Magnesium deficiencies also occur in mixtures that contain very large amounts of Ca or Cl. Symptoms of Mg deficiency occur first on the lower leaves. There is chlorosis of tissue between the veins, which remain green, and starting from the tips the blades die and usually curl upward. Purple color builds up on stems and petioles.

A plant in a pot may lose much of its color in a matter of weeks. You may first notice Mg symptoms at the top of the plant. The leaves in the growing shoot are lime-colored. In extreme cases, all the leaves turn practically white, with green veins. Iron deficiency looks much the same, but a sure indication of Mg deficiency is that a good portion of the leaf blades die and curl. Treat Mg symptoms with one-half teaspoon of epsom salts to each quart of water, and water as usual. The top leaves recover their green color within four days, and all but the most damaged should recover gradually. Continue to fertilize with epsom salts as needed until the plants are flowering well. If you are using soilless mixtures, include epsom salts regularly with the complete mixture. Because Mg deficiencies may indicate interference from other nutrients, foliar-spray with Mg to check your diagnosis if the plants are not obviously recovering.

Iron

Fe deficiency rarely occurs with indoor mixtures. Iron is naturally plentiful in most soils, and is most likely to be deficient when the soil is very acid or alkaline. Under these conditions, which sometimes occur in moist eastern soils outdoors, the iron becomes insoluble. Remedies include adjusting the pH before planting; addition of rusty water; or driving a nail into the stem. Commercial Fe preparations are also available. If the soil is acidic, use chelated iron, which is available to the plants under acidic conditions.

Symptoms of iron deficiency are usually distinct. Symptoms

appear first on the new growing shoots. The leaves are chlorotic between the veins, which remain dark green and stand out as a green network. To distinguish between Mg and Fe deficiencies, check the lower leaves for symptoms. Iron symptoms are usually most prominent on the growing shoots. Mg deficiencies will also show on the lower leaves. If many of the lower leaves have been spotting or dying, the deficiency is probably Mg. Mg deficiencies are much more common than iron deficiencies in marijuana.

Other Trace Elements

The following deficiencies are quite rare. Trace elements are needed in extremely small amounts, and often enough of them are present as impurities in fertilizers and water to allow normal growth. Many houseplant fertilizers contain trace elements. Trace-element deficiencies are more often caused by an extreme pH than by inadequate quantities in the soil. If a deficiency is suspected, foliar-spray with the trace element to remedy deficiencies. Our experience has been that trace-element deficiencies rarely occur indoors. We advise you not to add trace elements to indoor soils, which usually contain large amounts of trace elements already because of the addition of organic matter and liming compounds. It is easy to create toxic conditions by adding trace elements. Manufacturers also recommend using amounts of trace elements that may be too high for indoor gardens; so use them at about one-fourth of the manufacturer's recommended dose if an addition is found to be necessary.

Manganese   Mn deficiency appears as chlorotic and then necrotic spots of leaf tissue between the veins. They generally appear on the younger leaves, although spots may appear over the whole plant. Manganese deficiencies are not common. Manganese is present in many all-purpose fertilizers. Mn deficiencies may occur if large amounts of Mg are present.

Boron   B deficiency may occasionally occur in outdoor soils. The symptoms appear first at the growing shoots, which die and turn brown or gray. The shoots may appear "burned," and if the condition occurs indoors, you might think the lights have burned the plant. A sure sign of boron deficiency is that, once the growing tip dies, the lateral buds will start to grow but will also die. B deficiency can be corrected by application of boric acid, which is sold as an eyewash in any drugstore. Use one-fourth teaspoon per quart of water. Recovery occurs in a few days with healthy growth of new shoots.

Molybdenum   Mb deficiency occasionally occurs in outdoor soils, but rarely indoors. Mb is readily available at neutral or alkaline pH. Mb is essential for nitrogen metabolism in the plant, and symptoms can be masked for a while when N fertilizers are being used. Usually there is a yellowing of the leaves at the middle of the plant. Fertilizing with nitrogen may remedy some of the yellowing. However, Mb symptoms generally progress to the growing shoots and new leaves often are distorted or twisted. Mb is included in many all-purpose fertilizers.

Zinc   Zn-deficiency symptoms include chlorosis of leaf tissue between the veins. Chlorosis or white areas start at the leaf margins and tips. More definite symptoms are very small, new leaves which may also be twisted or curled radially. Zn deficiencies may occur in alkaline western soils. Galvanized nails can be buried or pushed into the stem. Commercial preparations of zinc are also available.

Copper   Cu deficiencies are rare; be careful not to confuse their symptoms with the symptoms of overfertilization. The symptoms appear first on the younger leaves, which become necrotic at the tips and margins. Leaves will appear somewhat limp, and in extreme cases the whole plant will wilt. Treat by foliar-spraying with a commercial fungicide such as $CuSO_4$.

SOILLESS MIXTURES

Soilless mixtures are an alternative to using large quantities of soil. Their main advantage is complete control over the nutrients that your plants receive. Soilless mixtures are also inexpensive and easy to prepare. They have a near-neutral pH and require no pH adjustment.

Soilless mixtures are made from soil components such as vermiculite, sand, or perlite. Soilless mixtures should be blended in such a way that they hold adequate water, but also drain well and do not become soggy. A good general formula is two parts vermiculite to one part perlite. About 10 percent coarse sand or gravel can be added to give weight and stability to the pots. Instead of vermiculite, you can use Jiffy-Mix, Metro-Mix, Ortho-Mix, Pro-Mix, and other commercial soilless mixtures, which are fortified with a small amount of necessary nutrients, including trace elements. You can also substitute coarse sand for perlite.

Potting

It is best to use solid containers with soilless mixtures rather than plastic bags. Grow the plants in one- to three-gallon containers. There won't be much difference in the size of the plants in one-gallon or in three-gallon sizes, but you will have to water a large plant every day in a one-gallon container. (The plants can always be transplanted to a larger container.) The pots must have drainage holes punched in the bottoms. Pot as usual, and add one tablespoon of dolomitic lime or two tablespoons of wood ash to each gallon of mixture.

Germinating

Plants may have problems germinating in soilless mixtures. The top layer of mixture often dries rapidly, and sprouts may die or not germinate. Young seedlings also seem to have difficulty absorbing certain nutrients (notably potassium), even though adequate amounts of nutrients are being added. Since this difficulty may retard growth, it is best to start the plants in small pots with soil. Use eight-ounce paper cups, tin cans, or quart milk containers cut in half. Mix three parts topsoil or potting soil to one part soilless mixture. Fill the starting pots and germinate as usual. When the plants are two to three weeks old, transplant to the soilless mixture. First moisten the soil, and then remove the soil as intact as possible. You might handle the transplant like making castles, by carefully sliding the moist soil out of the pot. Or you can cut away the sides of the container while you place the transplant in the soilless mixture. When watering, make sure you water around the stem to encourage roots to grow into the soilless mixture.

Peat pellets that expand are also good for starting seedlings. Plant several seeds in each pellet, and place it in the soilless mixture after the sprouts appear.

Fertilizing

Soilless mixtures can be treated with a trace-element solution. We have grown crops with no special addition of trace elements, and the plants completed their lives without showing symptoms of trace-element deficiency. In these cases there were apparently enough trace elements in the lime and the fertilizers that were used to provide the major nutrients. Many all-purpose fertilizers also contain trace elements. However, it is a good idea to treat soilless mixtures with a mild solution of trace elements before planting. Large plants can be treated a second time during the third or fourth month of growth. Do not use trace elements more often unless plants show definite trace-element deficiencies.

Iron is the only trace element that is needed in more than minute quantities. Iron can be supplied by mixing a few brads or nails into the soilless mixture.

Use any soluble fertilizer that is complete, that is, that contains some of each of the major nutrients. Choose one with a formula that is highest in N but contains a good portion of both P and K. For example, Rapid-Gro is 23-19-17 and works well for soilless mixtures.

Table 18 gives a formula that has worked well for us. The figures in it are a guide for estimating the amounts of fertilizer to use. When choosing a fertilizer by means of this chart, use N for a guide. For example, suppose the only fertilizer you can find that has good proportions of the major nutrients is a 20-15-15. Divide 5 (the figure for N in the table) by 20 (the figure for N in the fertilizer), and get the result 1/4. That is, the fertilizer is four times as concentrated in N as you need; so you would use one-fourth the amount of fertilizer shown in Table 18. For instance, during the vegetative stage, you would give the plants one-half to three-fourths of a level teaspoon of fertilizer per gallon of water each time you water.

It is also not necessary to fertilize in these ratios. You could use a 10-10-10 fertilizer throughout growth; you would use half the amounts listed in Table 18. The most important point is that the plants receive enough of each element, not that they receive specific proportions.

Fertilizing according to volume of fertilizer is not very accurate, and also does not take into account other variables (such as variety, light, temperature, etc.) that determine the amounts of nutrients your plants can use. However, it is a simple and useful way of estimating the plants' needs. You can more accurately gauge the plants' needs by giving a sample plant twice the con-

Table 18.

Guidelines for Fertilizing Soilless Mixtures

Growth Stage	N	P_2O_5	K_2O	Amount
Seedling	5	3	4	1.5 to 2 tsp/gal
Vegetative	5	2	3	2 to 3 tsp/gal
Flowering	5	5	3	0.5 to 1.5 tsp/gal

centration of fertilizer, and another half the concentration. Their performance will give you an idea of whether you are using too much or too little fertilizer. Too much fertilizer is the most damaging condition; so when in doubt give the plants less rather than more. Do not continue to give the plants the recommended amounts of fertilizer if the sample plant that is receiving less nutrients is growing as well as the other plants.

Another way of monitoring the plants' growth is to grow a few plants in a standard soil mixture. This will show you whether the plants in the soilless mixture are growing as fast as they should, and will give you a reference for diagnosing deficiencies.

Besides providing N, P, K, and the trace elements, you must also give your plants secondary nutrients. Ca is added by mixing a tablespoon of lime or two tablespoons of wood ash when preparing the soilless mixture. (Calcium is usually present in water and in many fertilizers as part of the salts that contain nutrients, for example, $Ca(NO_3)_2$.) Magnesium and sulfur are both found in common epsom salts, $MgSO_4$. Use one-eighth teaspoon of epsom salts to each teaspoon of 5 percent N. For example, if you are using a 20 percent N fertilizer, you would use half a teaspoon of $MgSO_4$ to each teaspoon of fertilizer. (Actually, enough sulfur is often present, either as part of the soilless mixture or as part of nutrient salts. to allow growth.) Magnesium can also be supplied by using dolomitic limestone.

Soilless mixtures are something between soil mixtures and water cultures (hydroponics). With hydroponics, the plants are grown in a tank of water. The fertilizers are added in solution, and the water solution is periodically circulated by a pump.

Another variation on soilless mixtures is to add a small amount of soil or humus to the soilless mixture. Some examples are:

1. 4 parts soilless mixture to 1 part soil;
2. 8 parts soilless mixture to 1 part humus;
3. 15 parts soilless mixture to 1 part limed manure.

Overfertilizing is less a problem with soilless mixtures than with soil, because higher concentrations of salts are tolerable in soilless mixtures and because excess salts are easily flushed out of the mixture. A good idea is to flush each pot once after two months of growth, again after four months. Any time the plants show symptoms of overfertilization, leach the pots immediately. Flood each pot with plain water so that it runs out the drainage holes. Continue flooding the pots until a couple of gallons of water have run through the pot. Don't fertilize for at least a week. Then fertilize with a more dilute solution than was used before.

Figure 51a. Over fertilization. Leaves usually turn bright gold and die, starting at the top of the plant. *Photo by Sandy Weinstein.*

Chapter Ten
Diseases and Plant Pests

Plants are considered diseased when their health or development is impaired enough that the adverse effects become visible to the eye. Disease may be caused by infectious microbes, such as bacteria or viruses, by pests such as insects, or by nutritional deficiencies or imbalances. However, for diseases that might affect your plants, there should be no need for a plant doctor. You'll be able to diagnose the symptoms after careful observation.

Leaves naturally drop from plants during the course of their lives. Not every leaf will develop perfectly or remain so. The small leaves that are formed during the first few weeks of growth normally die within three months. Leaves at the bottom of healthy plants often die because they are shielded by the upper leaves and are not receiving enough light to maintain life. For instance, in a garden receiving only 80 watts of fluorescent light, the plants may stay green only up to three or four feet away from the lights. Lower leaves may turn pale and yellow and then dry to gold or rust colors.

MICROBIAL DISEASES

Because *Cannabis* is not native to the Americas, most of the microbial diseases that attack the plant are not found in this country. Homegrown *Cannabis* is remarkably free of diseases caused by microbes, and there is little chance of your plants suffering from these diseases. Fungal stem and root rots seem to be the only ones of consequence. These occur only because of improper care. Watering too often, coupled with a stagnant, humid atmosphere, encourages stem rot to develop. Stem rot appears as a brown or black discoloration at the base of the stem and is soft or mushy to the touch. Allow the soil to dry between waterings, and be sure to water around the stem, not on it. Wipe as much of the fungus and soft tissue away as possible. If the rot doesn't disappear in a few weeks, treat it with a fungicide.

NUTRIENT DISEASES

Diseases due to nutrient deficiencies (see Chapter 9), are common indoors, and their symptoms usually worsen with time, affecting more and more of the plant. Whole leaves may be pale, or turn yellow or white; the condition may first afflict the bottom, or top, or the entire plant at once. Deficiency symptoms often appear as spots, splotches, or areas of chlorotic (lacking green) tissue. Sometimes necrotic (dead) tissue appears that is copper, brown, or gray. However, before you turn to Chapter 9, carefully inspect the plants for any signs of plant pests.

PLANT PESTS

The indoor garden is an artificial habitat where the plants live in isolation from the natural world. For this reason, few of you will have any problems with plant pests. However, indoor plants are particularly susceptible to pests once contaminated. In nature, the pest populations are kept in check by their natural enemies, as well as by wind, rain, and changing temperatures. Without these natural checks, pests can run rampant through the indoor garden.

The most common and destructive pests are spider mites and whiteflies. Spider mites are barely visible to the naked eye; they are ovoid-shaped. Juvenile mites are transparent and change to green as they suck the plant's tissue. Adults are tan, black, or semitransparent. False spider mites are bright red. Mites are usually well-established before you discover them, because they are so difficult to see.

Whiteflies are white (obviously) but look like tiny moths rather than flies. The adults are about 1/16 inch long, and you may not see one unless it flutters by the corner of your eye. Then shake the plants. If the result looks like a small snowstorm, the plants are infested with whiteflies.

Figure 52. *Left:* Spider mite (×16). *Right:* A match head dwarfs tiny spider mites.

The symptoms of infection by mites and whiteflies are similar. Symptoms usually appear on the lower leaves and gradually spread to the top of the plant. The first indications are that the plant loses vigor; lower leaves droop and may look pale. Look closely at the upper surfaces of the leaves for a white speckling against the green background. The speckles are due to the pests sucking the plant's chlorophyll-rich tissue. With time, the leaf loses all color and dies.

Pests are easiest to find on leaves that are beginning to show some damage. You can usually see mites and whitefly larvae as tiny dots by looking up at the lights through the undersides of the leaves.

To find out which pest you have, remove some damaged leaves and inspect the undersides under bright daylight. With spider mites, if you discover them early, a leaf may show only one or two tiny dots (adults) and a sprinkling of white powder (eggs) along the veins. In advanced cases, the undersides look dusty with the spider mites' webbing, or there may be webbing at the leaf nodes or where the leaflets meet the petioles. With whiteflies, you usually see the adults first. On the undersides of the leaves the whitefly larvae look like mites, but there is no

Figure 53. Mites appear as black specks when you look up to the lights from the undersides of the leaves. Also see Plate 14.

webbing, and there are tiny golden droplets of "honeydew" ex creted by the adult whiteflies.

Take quick action once you discover plant pests. If the plants are less than a month old, you will probably be better off to clean out the garden, in order to eliminate the source of the pests, and start over. As long as the plants are healthy they can withstand most attacks. The more mature the plants are, the less they are affected by pests. Whiteflies and mites sometimes disappear from flowering plants, particularly the female flowers. Mites are difficult to eliminate completely. Often a holding action will save a good crop.

If only a few plants in your garden are infected, remove them. Or else remove any leaves that show damage. If the plants are three or more months old, you might consider forcing them to flower while they are still healthy. Plants that are good-sized and still vigorous will usually stand up well to mites once they are flowering.

If you don't want to use insecticides, there are several alternative ways to keep the pests in check until flowering. Mix 1/8 to 1/4 pound of pure soap (such as Ivory flakes) thoroughly in one gallon of lukewarm water. Then cover each pot with foil or newspaper, invert it, and dip and swish the plant around

Figure 54.   Heavy whitefly infestation. Also see Plate 14.

several times in the soapy solution. Let it drip dry and rinse with clear water. Use the dunking procedure every week or two until the plants are larger. This is often enough to get the plants growing well and into flowering before the pest population can become a serious problem.

Two homemade sprays that can be effective are dormant oil sprays* and hot pepper sprays.

To make hot pepper spray, mix four hot peppers with one medium onion and one clove garlic.[213] Grind or chop and mash them along with some water. Cover the mash with water and allow it to stand a day or two. Add enough water to make two quarts. Strain through a coffee filter or paper towels in a funnel. Add one-half teaspoon of detergent and spray as you would an insecticide.

No one wants to use insecticides; yet they seem to be the only way to eliminate mites. There are a number of insecticides on the market that are relatively safe. Insecticides such as pyrethrum, rotenone, and malathion are relatively nontoxic to warm-blooded animals when used as directed. These are effective against many different plant pests besides mites and whiteflies. Additionally, they break down into harmless compounds such as carbon dioxide and water in a matter of days; so they do not persist in the environment.

Safe insecticides are used for vegetables. Follow all the package precautions. Do not use more, or more often, than recom-

---

*See "Insects and Pests" in the Outdoor Section.

mended. Overuse can kill the plant. The label will list the number of days to wait before you can safely ingest the plant, usually from two to 35 days after spraying.

Both mites and whiteflies generally complete their brief life cycles in about one to two weeks. Because sprays are not effective against the eggs, repeat the spraying about once a week for three successive weeks to completely eliminate the pests. Since their generations are short-lived, some pests may become resistant to the spray. This can be a problem with whiteflies. Try a different insecticide if the first one does not seem to be working.

Add a couple of drops of liquid detergent to each quart of insecticide solution. Detergent acts as a wetting agent and helps the insecticide to contact the pests and stick to the plant. Small plants can be dunked directly in the solution, the surest way to kill pests.

To spray the plants, start at the back of the garden so that you are working away from the plants already sprayed. Spray the entire plant and soil surfaces, paying special attention to the undersides of the leaves where pests tend to congregate. Stay out of the garden and keep the room closed that day.

Sulfur dusts can also be effective against mites and many other pests, and are safe to use. The easiest way to apply them is with a plastic "squeeze" bottle which has a tapered top. Make sure you dust the underside of the leaves.

Before using any insecticide, remove all damaged leaves. Do not use any insecticide during flowering. Rinse the plant with a clear water spray about one week after applying any insecticide, and once more before you harvest. Otherwise there may be residues left which will affect the taste of the grass.

There are several other pests that can be a problem, although they rarely seriously affect marijuana. Aphids are about 1/16 inch long and are black, green, red, or pink. They have roundish bodies with long legs and antennae. Some species have wings. They congregate on the undersides of leaves, which may then lose color and become curled or distorted. Aphids excrete honeydew droplets on the undersides of the leaves which can attract ants. If ants are also present, set out ant traps, because the ants will spread the aphids to other plants. A few successive washings in soapy water or one or two sprayings of the insecticides mentioned above should eliminate aphids.

Mealy bugs are white, about 3/16 of an inch long, and look like small, flat sowbugs. They don't seem to like marijuana and avoid it if other plants are present. Mealy bugs can be removed individually with cotton swabs and alcohol.

Gnats are attracted to moist soil that is rich in partially decayed organic matter such as manures. To discourage gnats when using manures, cover the top few inches in the pot with the soil mixture and no manure. Drench the soil with malathion solution for gnats or any other soil pest. Flypaper will also help against gnats as well as whiteflies.

Some people don't mind having a few pests on their plants. Whether you want to eliminate the pests completely or simply keep them in check may come down to whether you mind hearing the snap, crackle, and pop as their little bug bodies heat and explode when the harvest is smoked. Commercial marijuana, or any marijuana grown outdoors, will contain innumerable bugs and other small lifeforms.

Prevention

Whiteflies and spider mites are extremely contagious. Mites can be carried to the plant on hands, clothing, or an animal's fur. Many houseplant pests can fly or float to the garden through open windows. Mites crawl through cracks in walls and foundations during autumn, seeking warmth.

Many houseplants are popular because they can withstand abuse and infections by common plant pests. Your houseplants may harbor mites for years without your knowledge. You can find out if your houseplants have mites by placing some marijuana seedlings among the houseplants. Mites seem to enjoy young marijuana plants so much that the plants show symptoms of mites in a matter of weeks if any are nearby.

Hopefully, you'll never have to deal with pests. Prevention is the best policy. Use soil that has been pasteurized or sterilized to avoid bringing pest eggs and larvae into the garden. Keep the garden isolated from other plants. Use separate tools for the marijuana garden and for other plants. Screen windows in the garden with wire screen or mesh fabrics such as nylon.

Chapter Eleven
Maintenance and Restarting

To start a new crop, it is best to begin with a fresh soil. This is especially true if the plants were in small pots or were root-bound.

If you have fertilized regularly, the soil may contain near-toxic amounts of salts. Most of the salts build up in the top two-inch layer of soil. To salvage large quantities of soil, discard the top three-inch layer of soil from each pot. Add fresh soil and bulk fertilizers. Thoroughly mix and repot in clean containers.

It is generally not advisable to use the same soil for more than two crops. Although the used soil may not support healthy growth for potted plants, it is an excellent addition to any garden soil. Spread the soil as you would a mulch. The salt concentration is quickly diluted and benefits, rather than harms, garden soil.

Periodically clean the tubes and reflectors to remove dust and grime. As with windows, this dirt substantially decreases the amount of light the plants receive. Fluorescents lose approximately 20 to 40 percent of their original output within a year's use. Generally the higher-wattage tubes decline more rapidly than standard-output tubes. Vita-lite tubes last the longest, followed by standard fluorescents. Gro-tubes are the shortest-lived, and most growers replace them after two crops. Older tubes can be used to start seedlings and during the first month of growth. Since the plants are small and the light system is low, the old tubes generate enough light for healthy growth. Replace incandescent bulbs after 500 light hours.

PART 3.
Outdoor Cultivation

Chapter Twelve
Choosing a Site

There are several factors to consider when deciding where to plant, including sunlight, microclimate, availability of water, and condition of the soil. But the garden's security should be your first consideration. No matter what size your garden, rip-offs and confiscation are constant threats. But these risks can be minimized by careful planning and common sense.

In some counties, law-enforcement agencies take a tolerant attitude toward small gardens, and people grow *Cannabis* in their backyards. In other areas, police are not as enlightened and place an emphasis on cultivation busts. In either case, the larger the garden, the greater the potential danger.

Figure 55. A Nassau County police officer stands in a field of marijuana plants in Lattingtown, Long Island. *NYT Pictures.*

In Hawaii and California, where marijuana growing has become a booming business, helicopters have been a problem for commercial growers. Aircraft outfitted with visual or infrared equipment, dogs, and finks have all been used to seek out illicit plots. Aircraft equipment is least effective on steep slopes and where the vegetation is lush and varied. Where aircraft are a problem, growers prune marijuana to obscure its distinctive shape. The plants are difficult to detect from a distance when intercropped with bamboo, sunflowers, sugar cane, soybeans, or tall weeds (see Figure 60). Commercial growers often plant several small dispersed stands or many single plants, which are more difficult to detect and serve as insurance against total loss.

But rip-offs rather than the law are more of a problem for marijuana growers. From every section of the United States, reports confirm that marijuana theft has reached epidemic proportions, and even well-hidden plants fall prey to unscrupulous people. These lowlifes often search near hippie communities and popular planting areas. Their best ally is a loose lip; so keep your garden on a "need to know" basis.

WHERE TO GROW

Given the value of marijuana, many people think they'll grow an acre or two. But it is much harder to find spots suitable for large-scale farming than to find small garden plots. Large gardens require more planning and commitment, and usually a remote area. They may need a lot more time, energy, and investment in materials and labor-saving machinery than smaller gardens.

A small but well-cultivated garden, say, ten by ten feet, can yield over four pounds of grass each crop. By planning realistically, you'll harvest a good stash of potent grass rather than a lot of disappointment.

Most people who grow marijuana plant it in their backyards. They hide the plants from curious neighbors and passers-by with walls, fences, arbors, or similar enclosures. Some people plant Cannabis as part of their vegetable garden, pruning the plants to make them less conspicuous.

Gardeners often use ingenious ideas to keep their gardens secret. A woman on Long Island grows over thirty large plants in containers in her drained swimming pool. Although some of

Figure 56. Backyard garden in Massachusetts.

the plants reach a height of 12 feet, they can't be seen over the enclosing fence.

A couple living near Nashville, Tennessee, took the roof off their three-car garage and painted the walls white to create a high-walled garden. Other growers use sheds with translucent roofs.

Guerilla Farming

Many growers feel safer planting away from their property. Should the garden be discovered, they are not in jeopardy. On the negative side, they usually lose the close contact and control that a home gardener has.

Urban gardeners use makeshift greenhouses, rooftops, vacant lots, and city dumps. Vacant lots that are overgrown with lush weeds can support a good crop, if the marijuana plants get a head start on the indigenous weeds.

Fields, forest clearings, railroad rights-of-way, stream banks, runoff and irrigation ditches, clearings beneath high-tension lines, deserted farms and quarries, overgrown fields, and abandoned houses have all been used as garden spots. In areas where hemp is a problem weed, people plant seeds from high-potency marijuana in the same fields where the weedy hemp grows. Growers

Figure 57. Guerilla farm in the California Sierra.

harvest the plants in late July before they flower and before the fields are watched or destroyed by law enforcers.

Larger growers often look for rough, unpopulated terrain that is accessible only by plane, helicopter, four-wheel-drive vehicles, or long hikes. They avoid areas which hunters and hikers are likely to use before harvest.

Serious growers often find unusual places to start gardens. A grower in Chico, California, hacks through two hundred yards of dense underbrush and bramble to reach his clearing. In Oregon some growers maintain fields which are a grueling eight-hour uphill hike from the nearest road. Some Florida farmers commute to their island and peninsula gardens by boats. A master gardener in Colorado lowers himself by rope to a fertile plain 50 feet below a cliff.

A farmer in Hawaii wrote, "The main concern is to grow in an undetectable place where the plants can still get enough sun. This is becoming very difficult to find and some very elaborate subterfuges have been developed. People on Maui are growing plants suspended from trees and on tree platforms! Around here some people carry small plants in buckets far out on the lava fields where there is a light shading from Ohia trees and you don't leave tracks. Also people go into the sugarcane fields, tear out some cane, and put in their plants. I am sure many other things are being done."

LIGHT

Marijuana is a sun plant. The plants will grow in partially shaded areas, but about five hours of direct sun daily are needed for development into a lush bush. Marijuana does best when it has direct sunlight all day. If it grows at all in a heavily shaded area, it will be dwarfed and sparse—a shadow of its potential.

Try to choose a place that maximizes light. Flat areas get the most sunlight, but many growers prefer to use slopes and hillsides which help to hide the plants. Southern slopes usually receive more sun and stronger light than eastern and western slopes, which are shaded in the afternoon and morning, respectively. Northern slopes are rarely used, since they get the least sunlight and are also the coldest. Steeper slopes are shaded sooner than gradual slopes, and lower areas are shaded earlier than high ones.

Sunlight at high altitudes is more intense, because of the thinner atmosphere and the usually lower pollution. The atmosphere and pollutants at lower elevations absorb and scatter some of the solar radiation.

Backyard gardeners usually compromise between the need for maximum light and the need for subterfuge. An area that gets several hours of direct sunlight and bright unobstructed daylight for the rest of the day will do well. A garden exposed to the south usually gets the strongest light and is the warmest. Overhanging vegetation should be pruned so that the plants are shaded as little as possible.

Most marijuana strains are acclimated to tropical and semitropical latitudes, where the daytime is relatively short (10 to 14 hours, depending on season), but the sunlight is quite strong. At latitudes in the United States, the sun is not as intense (although in the summer the difference is small), but the days are longer, and the plants can grow extremely fast. It is not true that intense sunlight is needed to grow great marijuana. However, a summer characterized by clear sunny weather will usually produce a larger and slightly more potent crop than if the season is cloudy and rainy.

Sunlight can be maximized by adequate spacing and orientation of the garden. This is covered in Chapter 14.

Chapter Thirteen

Soil

Of all the factors involved in growing plants, soil is the most complex. It has its own ecology, which can be modified, enriched, or destroyed; the treatment it receives can ensure crop success or failure.

There is no such thing as the perfect soil for *Cannabis*. Each variety can grow within a wide range of soil conditions. Your goal is garden soil within the range for healthy growth: well-drained, high in available nutrients, and with a near neutral (7.0) *p*H. *Cannabis* grows poorly, if at all, in soils which are extremely compacted, have poor drainage, are low in fertility, or have an extreme *p*H.

There are several soil factors that are important to a grower; these include soil type, texture, *p*H, and nutrient content. We will begin this chapter by discussing each of these topics in succession, and will then turn to discussion of fertilizers, soil-preparation techniques, and guerilla farming methods.

TYPES OF SOIL

Each soil has its own unique properties. These properties determine how the soil and plants will interact. For our purposes, all soils can be classified as sands, silts, clays, mucks, and loams. Actually, soils are usually a combination of these ingredients. If you look carefully at a handful of soil, you may notice sand granules, pieces of organic matter, bits of clay, and fine silty material.

Sandy Soils

Sands are formed from ground or weathered rocks such as limestone, quartz, granite, and shale. Sandy soils may drain too well. Consequently, they may have trouble holding moisture and nutrients, which leach away with heavy rain or watering. Some sandy soils are fertile because they contain significant amounts (up to two percent) of organic matter, which also aids their water-holding capacity. Sandy soils are rich in potassium (K), magnesium (Mg), and trace elements, but are often too low in phosphorus (P) and especially nitrogen (N). N, which is the most soluble of the elements, is quickly leached from sandy soil. Vegetation on sands which is pale, yellowed, stunted, or scrawny indicates low nutrients, usually low N.

Sandy soils can be prepared for cultivation without much trouble. They must be cleared of ground cover and treated with humus, manure, or other N-containing fertilizers. In dry areas, or areas with a low water table, organic matter may be worked into the soil to increase water-holding capacity as well as fertility. Sandy soil does not usually have to be turned or tilled. Roots can penetrate it easily, and only the planting row need be hoed immediately before planting. Growers can fertilize with water-soluble mixes and treat a sandy soil almost like a hydroponic medium.

Sandy soils are also good candidates for a system of sheet composting (spreading layers of uncomposted vegetative matter over the garden), which allows nutrients to gradually leach into the soil layers. Sheet composting also prevents evaporation of soil water, since it functions as a mulch.

Silts

Silts are soils composed of minerals (usually quartz) and fine organic particles. To the casual eye, they look like a mucky clay when wet, and resemble dark sand or brittle clods when dry. They are the result of alluvial flooding, that is, are deposits from flooding rivers and lakes. Alluvial soils are usually found in the Midwest, in valleys, and along river plains. The Mississippi Delta is a fertile alluvial plain.

Silts hold moisture but drain well, are easy to work when moist, and are considered among the most fertile soils. They are frequently irrigated to extend the length of the growing season. Unless they have been depleted by faulty farming techniques, silts are rich in most nutrients. They often support healthy, vigorous vegetation. This indicates a good supply of N.

Mucks

Mucks are formed in areas with ample rainfall which supports dense vegetation. They are often very fertile, but may be quite acidic. They usually contain little potassium.

Mucks range from very dense to light sandy soils. The denser ones may need heavy tilling to ensure healthy root development, but the lighter ones may be cleared and planted in mounds. Mucks can support dense vegetation, and are often turned over so that the weeds thus destroyed form a green manure.

Clay Soils

Clays are composed of fine crystalline particles which have been formed by chemical reactions between minerals. Clays are sticky when wet, and can be molded or shaped. When dry, they form hard clods or a pattern of square cracks along the surface of the ground. Clays are usually hard to work and drain poorly. Marijuana roots have a hard time penetrating clay soils unless these soils are well-tilled to loosen them up. Additions of perlite, sand, compost, gypsum, manure, and fresh clippings help to keep the soil loose. Clay soils in low-lying areas, such as stream banks, may retain too much water, which will make the plants susceptible to root and stem rots. To prevent this, some growers construct mounds about six inches to one foot high, so that the stems and tap roots remain relatively dry.

Clay soils are often very fertile. How well marijuana does in clay soils usually depends on how well these soils drain. In certain areas "clay" soils regularly support corn or cotton. This type of soil will support a good crop of marijuana. Red color in clay soil (red dirt) indicates good aeration and a "loose" soil that drains well. Blue or gray clays have poor aeration and must be loosened in order to support healthy growth.

A typical schedule for preparing a heavy clay soil   In the late fall, before frost, turn soil, adding fresh soil conditioners, such as leaves, grass clippings, fresh manure, or tankage. Gypsum may also be added to loosen the soil. Spread a ground cover, such as clover, vetch, or rye. In early spring, as soon as the soil is dry enough to work, turn it once again, making sure to break up the large clods, and add composts and sand if needed. At planting time, till with a hoe where the seeds are to be planted.

As the composts and green manure raise the organic level in the soil, it becomes less dense. Each year, the soil is easier to work and easier for the roots to penetrate. After a few years, you may find that you only need to turn under the cover crop. No other tilling will be needed.

Loams

Loams are a combination of about 40 percent each of sand and silt, and about 20 percent clay. Organic loams have at least 20 percent organic matter. In actuality, a soil is almost always a combination of these components, and is described in terms of that combination, e.g., sandy silt, silty clay, sandy clay, or organic silty clay. Loams range from easily worked fertile soils to densely packed sod. Loams with large amounts of organic matter can support a good marijuana crop with little modification.

HUMUS AND COMPOSTS

Humus and composts are composed of decayed organic matter, such as plants, animal droppings, and microbes. Their nutrient contents vary according to their original ingredients, but they most certainly contain fungi and other microorganisms, insects, worms, and other life forms essential for the full conversion of nutrients. As part of their life processes, these organisms take insoluble chemicals and convert them to soluble forms, which plant roots can then absorb. Humus and composts hold water well and are often added to condition the soil. This conditioning results from the aerating properties and water-holding capacity of humus and composts, as well as balanced fertility.

Humus and composts have a rich, earthy smell, look dark brown to black, and may contain partially decayed matter, such as twigs or leaves. They are produced naturally as part of the soil's life process or can be "manufactured" at the site by gathering native vegetation into piles. Composts cure in one to three months, depending on both ingredients and conditions. Decomposition can be speeded up by turning and adding substances high in N. Composts are frequently acidic and are sweetened with lime when they are piled. This also shortens curing time, since the desirable microbes prefer a neutral medium.

TEXTURE

Soil texture refers to density, particle size, and stickiness, all of which affect the soil's drainage and water-holding characteristics. The most important quality of the soil for marijuana is that it drains well—that is, water does not stand in pools after a rain, and the soil is not constantly wet. In a well-drained soil, the roots are in contact with air as well as water.

Cannabis does best on medium-textured soils: soils that drain well, but can hold adequate water. Loams, silts, and sands usually drain well and are loose enough to permit good root development. Some clays and most mucks are too compact to permit the lateral roots to penetrate and grow. In addition, they often drain poorly, and when dry they may form hard crusts or clods, a condition marijuana cannot tolerate.

Several simple tests will indicate the consistency and drainage qualities of your soil. Test when the soil is moist but not wet. First, dig a hole three feet deep to check the soil profile. In a typical non-desert soil, you will find a layer of decaying matter on the surface, which evolves into a layer of topsoil. Most of the nutrients available to the plant are found at this level or are leached down from it. The topsoil layer is usually the darkest. It may only be an inch thick or may extend several feet. When in good condition, the topsoil is filled with life. Healthy topsoil contains abundant worms, bugs, and other little animals, and is interlaced with roots. If you can easily penetrate the underlying topsoil with your hands, its texture is light enough for healthy root growth.

The next layer, or subsoil, may be composed of a combination of sand, clay, and small rocks, or you may hit bedrock. Sandy, rocky, and loamy subsoils present no problems as long as the topsoil is at least six inches thick. Clay or bedrock often indicates drainage problems, especially if the spot has a high water table and stays wet.

Next scrape up a handful of soil from each layer. Press each handful in your fist, release it, and poke the clump with a finger. If it breaks apart easily, it is sandy or loamy. Clods that stick together, dent, or feel sticky indicate clay or muck.

To test for drainage, fill the hole with water. Wait half an hour to let the moisture penetrate the surrounding soil; then fill the hole with water again. If the water drains right through, you are working with a sandy soil. If it doesn't drain completely within 24 hours, the soil has poor drainage.

*p*H

The *p*H is a measure of how alkaline (bitter) or acid (sour) the soil is. The *p*H balance affects the solubility of nutrients, and helps the plant regulate its metabolism and nutrient uptake. The scale for measuring *p*H runs from 0 to 14, with 7 assigned as neutral. A *p*H below 7 is acid; a *p*H above 7 is alkaline.

Marijuana grows in soils with a *p*H range from 5 to 8.5, but it thrives in nearly neutral soils. Relative to other field crops, it has high lime requirements, similar to those for red or white clover or sunflower. But it does well in fields where plants with medium lime requirements, such as corn, wheat, and peanuts, are grown.

The solubility of nutrients is affected by soil type as well as by the *p*H. In soils with a high content of organic matter, all nutrients are soluble between 5.0 and 6.5. Phosphorous, manganese, and boron are less soluble at *p*H values above 6.5. Acid soils are usually found in the United States east of the 100th meridian and along parts of the West Coast. These areas have ample rainfall, dense plant growth, and a deep topsoil layer. Marijuana does best in acid soils when the *p*H is adjusted to a range of 6.3 to 7.0.

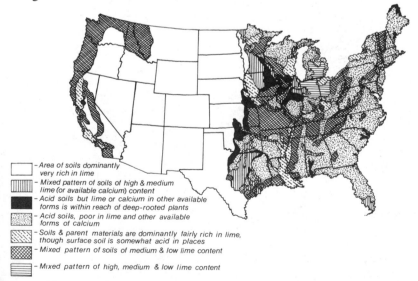

- Area of soils dominantly very rich in lime
- Mixed pattern of soils of high & medium lime (or available calcium) content
- Acid soils but lime or calcium in other available forms is within reach of deep-rooted plants
- Acid soils, poor in lime and other available forms of calcium
- Soils & parent materials are dominantly fairly rich in lime, though surface soil is somewhat acid in places
- Mixed pattern of soils of medium & low lime content
- Mixed pattern of high, medium & low lime content

Figure 58. Map of *p*H. Adapted from *Soils and Fertility*.[227] *Map by E.N. Laincz.*

Mineral soils in the dry western states may be slightly acid to highly alkaline. Most nutrients are very soluble in these soils, as long as the pH ranges from 6.0 to 7.5. Some of these soils are too alkaline (over 8.5); so their pH must be adjusted to near neutral to ensure healthy growth.

Adjusting the pH

First test the soil pH in the garden area. Previous gardeners may have adjusted native soils, or your yard soil may have been trucked in to cover poor native soils, so that the pH of your garden soil may be different from that of other soils in the area. Different soils vary in the amount of material needed to adjust the pH. Sandy soils do not require as much as loam, and loam requires less than clays, partly because of the chemistry, and partly because of the density and physical qualities of the soils' particles.

Adjusting Acid Soils

Acidic soils are treated with limestone, which is expressed as an equivalent of calcium carbonate ($CaCO_3$). Limestone is usually quarried and powdered, contains large amounts of trace elements, and comes in different chemical forms: ground limestone, quicklime, and hydrated lime (which is the fastest-acting form). Dolomitic limestone is high in magnesium and is often used to adjust magnesium-deficient soils, such as those found in New England. Marl (ground seashells) is also mostly lime and is used to raise soil pH. Eggshells are another source of lime. They should be powdered as finely as possible, but even so, they take a long time to affect the soil. Wood ashes are alkaline and very soluble; so they have an almost immediate effect.

Every commercial lime has a calcium carbonate equivalent or neutralizing power which is listed on the package. To find out how much to use, divide the total amount of limestone required by the pH test (see Figure 59) by the calcium carbonate equivalent. For instance, a field requires fifty pounds of limestone, but the calcic limestone you are using has an equivalent of 1.78. Divide the 50 by 1.78. The resulting figure, about 29 pounds, is the amount required. Commercial limes also list the grade or particle size of the powder. In order of fineness they are: superfine, pulverized, agricultural grade, and fine meal. The finer the grade, the faster the action.

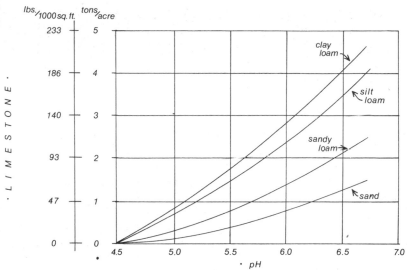

Figure 59. Approximate amount of lime required to adjust *p*H of a 7" layer of different types of soil. Adapted from USDA Bulletin No. 2124, 1959. *Graph by E.N. Laincz.*

For best results, lime should be added at least four or five months before planting. In this way, the lime has a chance to react with the soil. But acid soils can be limed profitably any time before planting, or after, as long as the lime does not come into direct contact with the plants. Most growers add lime at the same time that they fertilize and turn the soil. That way, tilling and conditioning are handled in one operation. The lime should be worked into the soil to a depth of ten inches. Lime can also be added by spreading it before a rain. Make sure that the soil is moist enough to absorb the rain, so that the lime does not run off. Growers who have not adjusted the *p*H can dissolve lime in water before they irrigate. However, this is not advised if the water runs through a hose or pump, because mineral buildup may occur in the equipment.

Adjusting Alkaline Soils

Most alkaline soils have a *p*H no higher than 7.5, which is within the range for optimum growth. Soils that are too alkaline can be adjusted by adding gypsum, which frees insoluble salts, and sulphur, which neutralizes insoluble salts. Sulphur compounds include iron, magnesium, and aluminum sulphate. Marijuana has a low tolerance for aluminum; so marijuana growers should use

Table 19.

Recommended Amounts of Sulfur (in Pounds) to Lower Soil *p*H to Approximately 7.0[a]

Initial *p*H of Soil	Sand Soils		Clay Soils	
	1,000 sq.ft.	Acre	1,000 sq.ft	Acre
7.5	6	250	10.5	450
8.0	23	1,000	29	1,250
8.5	35	1,500	35	1,500
9.0	51	2,200	Not recommended[b]	

[a]These amounts are for broadcast application; when sulfur is turned into the soil, use .75 of these amounts.
[b]Use techniques for alkali soils.

iron or magnesium sulphate in preference to aluminum sulphate. Sulphur and gypsum are worked into the soil in the same manner as lime.

Some growers correct alkaline soils by adding an organic mulch or by working acidic material into the soil. Cottonseed meal, which is acidic and high in nitrogen, can also be used. As it breaks down, cottonseed meal neutralizes the soil. Pine needles, citrus rinds, and coffee grounds are all very acidic, and can be used to correct alkaline conditions. The addition of soluble nitrogen fertilizers aids the breakdown of these low-nitrogen additives. (See Table 22 in the section on "Fertilizers" in this chapter.)

Adjusting Alkali Soils

Alkali soils (*p*H usually above 8.5) are hardpacked and crusty, and sometimes have an accumulation of white powdery salts at the surface. They may not absorb water easily and can be extremely difficult to work. To prepare alkali soils with a permeable subsurface for cultivation, farmers leach them of their toxic accumulation of salts. The soil is thoroughly moistened so that it absorbs water. Then it is flooded so that the salts travel downward out of contact with the roots. Gypsum can be added to free some of the salts so that they leach out more easily. Gypsum can be added at the rate of 75 lbs per 100 sq.ft., or 18 tons per acre. Leaching requires enormous quantities of water, an efficient irrigation system, and several months.

(Text continued after color section.)

Plate 1. Skylights are a good source of bright, unobstructed light. Thai plant (closest) and Colombian plants reached over 14 feet in six months (San Francisco).

Plate 2. *Top:* A hidden garden using fluorescent light, foil reflectors, and bag containers. Plants are ten weeks old. *Bottom:* Simple-to-construct dome greenhouse in southern California. At two months, some of these plants are six feet tall.

Plate 3. *Upper left:* Stem of a female plant. *Upper right:* In full sunlight, a pruned plant can grow incredibly dense. *Bottom:* A garden in the wilds of Oregon mountains.

Plate 4. Marijuana does well in most gar
Top: Here a female plant is in early b
at five months. The main stem was cli
at three months (Berkeley). *Middle:* L
branches are spread out to catch the
Bottom: A female bud about two week
fore harvest. Leaves show some damage
leafhoppers (insect shown).

Plate 5. A giant sinsemilla cola grown from Mexican seed in northern California.
Photo by A. Karger.

Plate 6. *Top:* Purple colors often appear late in life, when vigor is waning. *Lower left:* Resin glands glistening on a purple, female flowering shoot. *Lower right:* Yellow male flowers and purple leaves against a normal green leaf.

Plate 7. *Top:* Male flowers at different stages in development. A line of resin glands can be seen on the anthers of the open flowers. *Lower left:* Resin glands form a line on opposite sides of each anther. (x 40). *Middle right:* Male flowers in full bloom. The leaves are covered with fallen pollen. *Photo by Sandy Weinstein. Lower right:* Gland heads (two crystal globes) may fall with the pollen grains. Mature grains are spherical in field of focus (x 40).

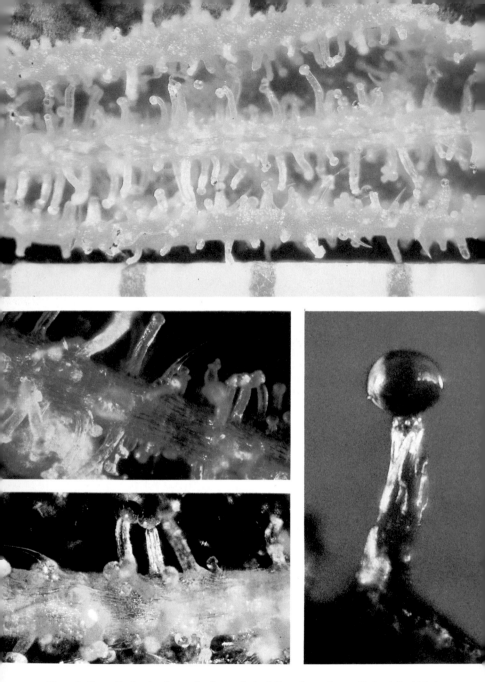

Plate 8. *Top:* Resin glands on the lower (adaxial) surface of a small, fresh leaf blade. Integrals are one millimeter (× 16). *Middle and lower left:* Stalked glands are concentrated along the veins of the lower leaf surface (× 40). *Lower right:* (× 100).

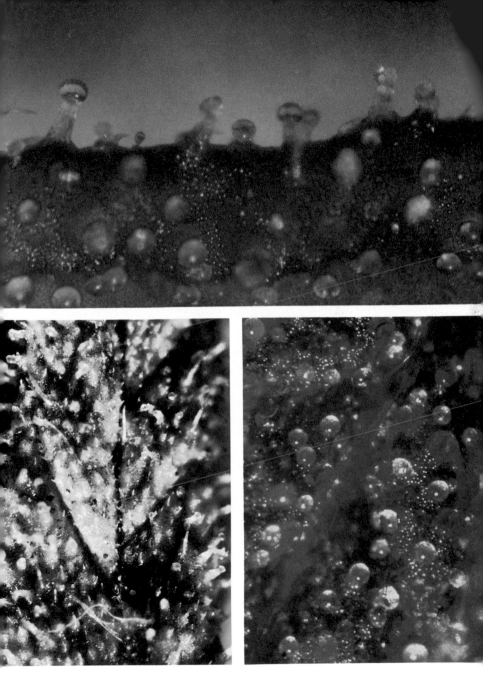

Plate 9. *Top:* Upper (abaxial) fresh leaf surface. Left of picture, from left to right: Sharp-pointed cystolith hair, stalked gland, and tiny bulbous gland (x 40). *Lower left:* Upper surface of a Thai leaf (x 16). *Lower right:* Upper surface of fresh home-grown Colombian leaf (x 40).

Plate 10. A young female flower (homegrown Colombian). Resin glands are not yet fully developed (× 16).

Plate 11. *Top left:* A mature female flower from the same plant as in Plate 10. The flower bract is swollen from the ripe seed it contains. Notice the well-developed resin glands (× 25). *Top right:* A mixture of seeds from common marijuana varieties shows comparative size. *Bottom:* The tip of a sinsemilla flower at harvest. Notice cream-colored stigmas to the left and the fresh, clear resin glands (× 40).

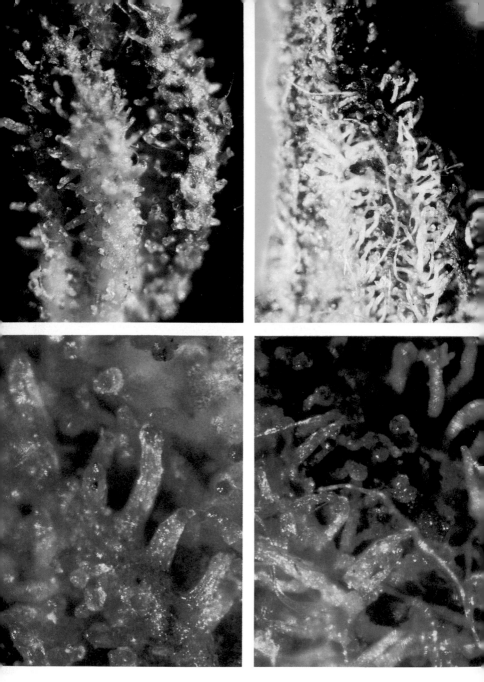

Plate 12. *Upper and lower left:* An overly ripe sinsemilla flower bract. Many gland heads are brown or missing (top, × 16; bottom, × 40). *Upper and lower right:* Carefully handled Thai weed with intact glands. Notice the high concentration of glands and very long stalks on this bract (top, × 16; bottom, × 40).

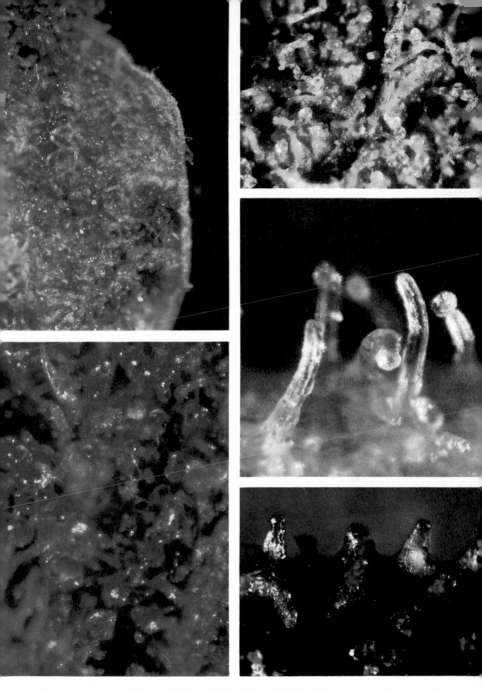

Plate 13. *Upper and lower left:* A Colombian gold. Gland contents are brown and stalks have deteriorated on this bract (top, × 16; bottom, × 40). *Top right:* Hawaiian; well-handled and showing little deterioration (bract × 40). *Middle right:* Gland heads easily detach from stalks when overripe (leaf vein × 40). *Lower right:* Stalked glands on both upper and lower leaf surfaces beginning to brown (leaf margin × 40).

Plate 14. *Top:* Whitefly larvae and their honeydew excretions on the lower surface of of a leaf. *Middle left:* Leaf showing whitefly damage and a tiny adult. *Lower left:* White speckles on leaves indicating mite damage. *Lower right:* An overdose, or overuse of pesticide, can kill the plant.

Plate 15. *Upper left:* Healthy green plant next to a N-deficient plant. *Middle left:* Ultraviolet burn. Plant was moved outdoors without conditioning. *Lower left:* "Bonsai" marijuana grown from a cutting. *Upper right:* Mg-deficient plant has chlorotic leaves dying from their tips. *Lower right:* Afghani variety, with characteristically wide leaf blades, shows minor symptoms of N deficiency (pale leaves and red petioles).

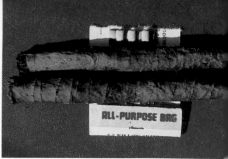

Plate 16. *Upper left:* Male flowers lose some green and turn "blond" during slow drying. *Upper right:* Cigar joints made with undried marijuana, which is wrapped with long blades of fan leaves before drying. *Bottom:* Sequence shows change in color in one day from sun curing.

Another method of reclaiming alkali soils is by adding a thick mulch and letting it interact with the soil during the winter. The mulch should be about nine inches thick, or 130 lbs or more per 100 sq.ft. This thick layer neutralizes the salts and also helps the soil to retain moisture.

NUTRIENTS

Marijuana is a high-energy plant which grows quickly to its full potential in a fertile soil that is rich in available nutrients. Nutrients are found in the soil's parent materials: sand, clay, humus, minerals, rocks, and water. Nutrients dissolve in soil water (soil solution), which is then absorbed by the plant. In complex chemical processes, roots release ions in exchange for nutrients that are dissolved in the soil solution.

The soil acts as a reservoir for the nutrients. Most of them are in non-exchangeable forms: that is, they do not dissolve, or dissolve only slightly in water. Only a small percentage of the total reserve is free at any time as the result of chemical processes or microbial action. Healthy soils maintain a balance between free and unavailable nutrients, so that the plants they support continually receive the right amounts of required nutrients. Alkali soils have large supplies of compounds which are extremely soluble. The solution is so concentrated that alkali soils are often toxic to plants.

There are three primary nutrients, N (nitrogen), P (phosphorus), and K (potassium). These are the nutrients that gardeners are most likely to be concerned with and which most fertilizers supply. Soils are most likely to be deficient in one of these nutrients, especially N.

In addition to the primary nutrients, soil supplies plants with three secondary nutrients, Ca (calcium), Mg (magnesium), and S (sulfur), and seven micronutrients: iron, boron, chlorine, manganese, copper, zinc, and molybdenum. Although deficiencies of all the secondary and micronutrients are reported from various parts of the United States, serious deficiencies do not occur often.*

Marijuana absorbs nutrients primarily through a fine network of lateral roots which grow from the taproot. Lateral roots

---

*For a discussion of the symptoms of nutrient deficiencies in marijuana, see Chapter 9.

may spread over an area with a diameter of five feet, and may go as deep as the roots can penetrate. Plants in deep sandy soils or in soils that have porous mineral subsoils may grow roots as deep as seven feet. Roots which can absorb nutrients from a larger area are more likely to fulfill the plants' needs than are shallow roots which result in shallow topsoil layers over compacted subsoils. When the roots have a large area from which to absorb nutrients, the soil does not need to be as fertile as when the roots are restricted to a small area by poor soil or by being grown in pots.

You can get a good indication of soil fertility by observing the vegetation that the soil supports. If the vegetation is varied, has a lush look to it, is deep green, and looks vigorous, it is probably well-supplied with nutrients. If the plants look pale, yellowed, spindly, weak, or generally unhealthy, the soil is probably deficient in one or more nutrients.

Testing

Agricultural colleges, County Extension Agents, and private companies perform soil analyses for a small fee from a sample you mail to them. The tests include nutrients, pH, and texture analyses, and are very accurate. There are also simple-to-use test kits available at nurseries and garden shops which give a fair indication of soil fertility and pH. Test results include a suggested fertilizer and lime program catered to the soil's individual requirements for the crop to be planted. Marijuana has nutrient requirements similar to those for corn, wheat, and sugarcane, and prefers just a little more lime (a more alkaline soil) than those crops; so soil can be fertilized as it would be for those crops.

Soil tests are one indication of soil fertility. They test for available nutrients, but not for reserves that are held in the soil. Test results may also vary because of recent rainfall, changes of moisture content, and seasonal changes. Most soil tests do not measure the ability of the soil to make nutrients available. This is a very important factor when considering a fertilizer program and should not be overlooked. As an example, an uncultivated field showed only moderate amounts of N available, and indicated a need for N fertilizer. The vegetation—tall grass, weeds, and bush—had a healthy look and was dark green, and the lower leaves remained healthy. Obviously, the soil was able to supply

Figure 60. Marijuana is often interspersed with corn. *Photo by K. Spangenberg.*

an adequate amount of N to the plants, which withdrew it from the soil solution as it became available. The soil and plants had reached a balance, and the soil solution slowly became more dilute over the course of the season.

To a great extent, the soil's ability to maintain a constant and adequate supply of nutrients depends on the soil's humus content. Humus can support dense populations of microorganisms. As part of their life processes, microorganisms decompose organic matter in the humus. Nutrients contained in the organic matter are released by microbes as simple inorganic molecules (e.g., NO_3) which can dissolve in soil water. Generally, soils with a high humus content can keep plants supplied with more nutrients than soil tests indicate.

The Primary Nutrients

If you look at any fertilizer package, you will note three numbers on the package. They stand for N-P-K, always in that order. Marijuana does best in a soil which supplies high amounts of N and medium amounts of P and K.

Nitrogen   The availability of N is the factor most likely to limit the growth of marijuana. For fast healthy growth, marijuana requires a soil rich in available N. Nitrogen is constantly being replaced in the soil solution by microbial breakdown of organic matter. Some microorganisms can use N directly from the atmosphere. They release N as waste in the form NO_3, which is the primary form in which plants absorb N. A small amount of N is also dissolved in falling rainwater. When the soil is moist, it loses N through leaching and to plants. In its available form (NO_3, NO_2, NH_4), N is very soluble and may be carried away with runoff or may drain into the subsoil.

Probably the most accurate method of measuring a soil's ability to produce available N is by the percentage of organic matter in the soil (see Table 20). Organic matter releases N at a rate that is determined by the type of soil, the temperature, and the moisture. Generally, the more aerated and warmer the soil, the faster organic matter decomposes and releases N. Most professional testing services report the percentage of organic matter, and some sophisticated kits can also test for it.

Table 20.

Nitrogen Released from the Soil During the Growing Season

Percentage of Organic Material in Soil	Released per Acre (per 1,000 sq.ft.)		
	Sandy Loam	Silt Loam	Clay Loam
1%	50 lb (1 lb, 2 oz)	20 lb (7.5 oz)	15 lb (5.5 oz)
2	100 lb (2 lb, 5 oz)	45 lb (1 lb, 1 oz)	40 lb (15 oz)
3		68 lb (1 lb, 9 oz)	45 lb (1 lb, 1 oz)
4		90 lb (2 lb, 1.5 oz)	75 lb (2 lb, 12 oz)
5		110 lb (2 lb, 9 oz)	90 lb (2 lb, 1.5 oz)

In its available state, N is tested in two compounds, ammonium (NH_4) and nitrate (NO_3). Test results are converted into PPM (parts per million) of N and then added to arrive at the total amount of N available in the soil. The formulas to convert nitrate and ammonium to N are (NO_3) X .226 = N, (NH_4) X .78 = N. Each PPM indicates 10.7 pounds of N per acre available in the top 7.87 inches. If the soil level is deeper, there is probably more N available. If it is shallower, less is available. But a test for available N gives only a fair approximation of the soil's ability to feed the plant. An individual test may be untypical because of recent leaching or depletion during the growing season.

An intensively cultivated crop of hemp takes about 250 pounds of N per acre or six pounds per 1,000 square feet from the soil during the growing season. When the plants are spaced well apart, the crop does not require as much N.

Fields which have more than 200 lbs of available N per acre (or 4.5 lbs per 1,000 sq.ft.) at the start of the growing season require no additional fertilization. Soils with less available N will probably yield a larger crop if they are given additional N. Actually, the amount of N that can profitably be used depends on the soil and its potential to produce N as well as on other factors: how fast N is lost, the soil depth, and moisture content.

One way to calculate the amount of N to add to the soil is to build your soil to an "ideal" level. For example, an Iowa silt loam may test about 1.6 pounds of N per 1,000 sq.ft. and an organic content of 3 percent. Together, the available and potential N total about 3.2 lbs per 1,000 sq.ft. To increase the available N to 4.5 lbs per 1,000 sq.ft., you would need to add 1.3 lbs of N.

Phosphorus   P is an important nutrient which is used directly by the soil bacteria as well as by the plant, so that an increase in the amount of P in the soil often results in an increase of N. Because of P's low solubility, it is rarely leached from the soil. It is usually found in the greatest concentration in the soil's top layers, where it accumulates as a result of decomposition of organic matter.

In slightly acid organic soils, up to one percent of the total P is available at any time. The total amounts of P in soils range from 1,000 to 10,000 lbs per acre. For example, a typical Kansas prairie soil has 3,000 lbs per acre. In soils with a lower *p*H, more of the P is tied up in insoluble compounds of iron or aluminum. In highly alkaline soils, the P forms insoluble compounds with calcium.

Insoluble P reacts with the dilute acids that are released during the decomposition of organic matter. These compounds are available to the plants. Both the chemical processes in which P is released and the organic processes of decomposition occur faster in warm soils.

If P is available, young plants absorb it rapidly, and may take in 50 percent of their lifetime intake by the time they are only 25 percent of their adult size. Young plants grown outdoors in cold weather may grow slowly until the soil warms up and more P is available. Older plants grown out of season in cold weather sometimes exhibit purple leaves. This condition may result from a P deficiency, because of the unavailability of P at low temperatures.

Most soil-test kits test available P, but the nutrient value of P is usually expressed as phosphoric acid (P_2O_5), which is converted using the formulas $P \times 2.3 = (P_2O_5)$, (P_2O_5) divided by 2.3 = P. Any soil that has available P of 25 lbs per acre (.58 lbs per 1,000 sq.ft.) or more is well-supplied with P. Stated in terms of phosphoric acid, this is $25 \times 2.3 = 57.5$ lbs per acre (1.33 lbs per 1,000 sq.ft.).

Most inexpensive soil kits test *available* P. Soil that tests less than 1 PPM or 10.7 lbs per acre (.25 lbs per 1,000 sq.ft.) of available P should be tested to make sure there are adequate reserves, or can be fertilized to assure maximum yield. Soil-test kits give only a fair indication of the P available. A low reading may indicate that the plants are absorbing P as fast as it breaks down from its unavailable form, especially during early growth! The main factors affecting the rate at which P becomes available are the total amount of reserve P in the soil and the pH.

Most professional soil analyses include a report of reserve P. Generally soils with reserve P of 3,000 lbs per acre (70 lbs per 1,000 sq.ft.) do not need additional P. Intensively cultivated and cropped fields may have had their reserve supply depleted, and will lock up available P that is supplied as fertilizer until a balance is reached.

Potassium K is found in adequate quantities in most soils which have a pH within the range needed for growing marijuana. K is held in soils in three forms: unavailable, fixed, and readily available. Most K is held in the unavailable form as part of the minerals feldspar and mica. But a small percentage of the total K in any soil is held in fixed, slightly soluble forms. Some of these can be absorbed and used directly by the plant. The exchangeable K is equal to a fraction of the fixed K. Each soil maintains a balance or ratio of unavailable to fixed and to exchangeable forms. Organic soils have a higher percentage of K in the fixed or available form than mineral soils. As K is used by the plants, some of the unavailable K goes into the more available forms. Plants can use K in both the soluble and the fixed forms.

Most clays and soils that are well-limed have adequate reserves of K. Acidic soils generally have low K reserves. Mucks, silts, and peats have low reserves of K, and have little capacity to hold it chemically when it is applied. Sands have K reserves, but little capacity to convert it to a fixed or available form. Most western soils have adequate reserves of K. The exchangeable K in soils becomes fixed if the soil dries out; so the available K of a recently dried soil is usually low.

K is tested in its elementary state, but when described as a nutrient, it is given as potash (K_2O). The formulas for conversion are K $\times$ 1.2 = (K_2O), (K_2O) divided by 1.2 = K. Soils with 180 lbs or more of available potash per acre (4 lbs per 1,000 sq.ft.) have an adequate supply. The total reserve K should test no lower than 900 lbs per acre (21 lbs per 1,000 sq.ft.).

The Secondary Nutrients

Magnesium (Mg), calcium (Ca), and sulfur (S) are usually found in adequate quantities in soils suitable for growing marijuana. However, some New England soils do have Mg deficiencies. Soils which have a neutral or near-neutral pH almost always have adequate Ca and sulfur levels.

Magnesium deficiencies are corrected by adding 50 to 100 lbs of Mg per acre (2.25 lbs per 1,000 sq.ft.). The most inexpensive way to add Mg is to use dolomitic limestone for adjusting soil pH. Dolomitic limestone is about 12 percent Mg (see Table 21); so 800 lbs of it are needed to supply 100 lbs of Mg. Dolomitic limestone releases Mg to the soil gradually. For faster action, epsom salts (magnesium sulfate, $MgSO_4$) can be used. Five hundred lbs of epsom salts are required to supply soil with 100 lbs of Mg. Mg deficiencies can also be corrected by using foliar sprays. Dissolve one ounce of epsom salts in a gallon of water and spray all foliage.

Table 21.

Nutrient Content of Agricultural Limestone[a]

Nutrient	Amount	
Calcium carbonate	75.4%	
Magnesium carbonate	17.1%	
Iron	0.4%	
Potassium	0.2%	
Sulfur	0.1%	
Manganese	330	PPM
Phosphorus	210	PPM
Zinc	31	PPM
Boron	4	PPM
Copper	2.7	PPM
Molybdenum	1.1	PPM

[a] Average from results of 220 samples from 35 states. From P. Chichilo and C.W. Whittaker.[203]

Micronutrients

Micronutrients are used by plants in minute quantities, and most soils contain enough of them to meet plant requirements. Home gardeners and guerilla farmers seldom encounter any micronutrient deficiencies. But heavily cropped lands sometimes develop a deficiency of one or more micronutrients because of crop depletion. Micronutrients are made available to the plants only if there is a delicate balance in the soil chemistry, and it is easy to create toxic conditions by adding them to soil when they are not needed. For that reason, soils should be treated with micronutrients only when symptoms occur or when the deficiency is known by analysis or past experience. Only small quantities of additives are required for treatment. Manures, composts, other organic fertilizers, lime, rock powders, and ash contain large quantities of trace elements. Active organic additives quickly release micronutrients in a form that is available to the plants.

Boron   Boron deficiencies in marijuana occur in acid soils as a result of depletion by heavy cropping. The areas most affected by it are vegetable fields in the mid-Atlantic states, alfalfa and clover fields east of the Mississippi, and truck farms and orchards in the Northwest. Boron is found in phosphate fertilizers, gypsum, and lime, and is the main ingredient of boric acid and borax. When borax or boric acid are used, they are applied at the rate of 10 to 20 lbs per acre. They are used as a foliar spray at the rate of 1 ounce per gallon of water.

Chlorine Chlorine deficiency does not normally occur. Some chemical fertilizers contain chlorine, and toxic conditions occur infrequently. Toxic chlorine conditions are eliminated by leaching.

Copper Copper deficiencies occur infrequently in truck farms in Florida, California, and the Great Lakes region. Wood shavings and tobacco contain large amounts of copper. A foliar spray composed of 1 ounce each of calcium hydroxide and copper sulfate (a fungicide) per gallon of water is used by commercial vegetable growers.

Iron Iron deficiencies occur in orchards west of the Mississippi and in Florida, and in alkaline soils in which iron is largely insoluble. Lowering soil pH often solves the problem. Chelated iron, which is water-soluble, is available at most nurseries and quickly supplies iron even when pH is extreme. Humus and seaweed are excellent sources of iron.

Manganese Manganese deficiencies occur in the Atlantic states, the Great Lakes area, Utah, and Arizona. Manganese is found in manure, seaweed, and some forest leaf mold (especially hickory and white oak). Manganese deficiencies can be corrected by using a foliar spray of manganese sulfate at the rate of 0.5 to 1.0 oz. per gal. Soil is sometimes treated with manganese sulfate at the rate of 20 to 100 lbs per acre. In neutral or alkaline soils, most of the manganese sulfate becomes fixed and unavailable to the plants by the end of the growing season.

Molybdenum Molybdenum deficiencies occur primarily along the Atlantic and Gulf coasts and in the Great Lakes region. Plants need extremely small amounts of molybdenum, less than 1 PPM in leaf and stem tissue. Molybdenum deficiencies occur when the soil is too acidic. By raising the pH level, one can make molybdenum available.

Zinc Zinc deficiencies occur in soils throughout the U.S., primarily because of heavy cropping. It is most likely to occur in acid-leached sandy soils, and in neutral and alkaline soils where it is insoluble. In soils with high amounts of available P, zinc is also unavailable. Many deciduous tree leaves and twigs, composts, slag, and rock phosphate contain large amounts of zinc. Zinc sulfate is used as a foliar spray at the rate of 3 oz. of zinc sulfate per gallon of water, or as a soil treatment at the rate of 100 lbs per acre. Some orchard growers drive galvanized nails into the trees to provide zinc.

FERTILIZERS

Most soils can benefit from a realistic soil-conditioning program. Most organic programs build soil, and minimize leaching and runoff. Programs using chemical fertilizers emphasize immediate increases in yield and a minimum of labor. The approach that you use should be tailored to the soil's needs and to your situation and goals. For example, a home gardener interested in building soil quality can easily add manure or compost to his garden. But a guerilla farmer may use concentrated chemical fertilizers, which are easy to transport to a remote area. A farmer cannot use the labor-intensive techniques which a small planter might use as a hobby. Many gardeners use both organic and inorganic fertilizers.

Organic Fertilizers

Organic fertilizers are usually less concentrated than chemical mixes. Their bulk consists of fibrous materials which condition the soil by aiding drainage and increasing the organic content and water-holding capacity. As they are decomposed by microbial action, the nutrients they contain are released in soluble form. Since this is a gradual process, there is little chance of creating toxic conditions.

Manures and composts are basic, all-purpose conditioners. They contain adequate amounts of most of the nutrients that marijuana absorbs from the soil and can be used generously. Uncomposted manures are "active" and should be used only in the fall. Over the winter they compost in the ground. Composts and composted manures can be added in the spring. Table 22 lists some common organic fertilizers which are usually available. Some of them, such as bone meal and granite dust, break down

Table 22.

Some Common Organic Fertilizers

Fertilizer	Percentage[a] of			Remarks
	N	P_2O_5[b]	K_2O[b]	
Bloodmeal	15.0	1.30	.70	N readily available.
Bone meal	4.0	21.0	.20	Releases nutrients slowly.
Cow manure[c]	.29	.17	.35	Fresh, 80 percent water. Excellent soil conditioner; apply in winter or compost for use in spring. Medium availability.
Coffee grounds	2.0	.36	.67	Highly acidic; best for use in alkaline soils.
Corn stalks	.75	.40	.90	Break down slowly; chopped stalks make excellent conditioner for compact or dense soils.
Cottonseed meal	7.0	2.5	1.5	Highly acidic; nutrients become available over the course of the growing season.
Dried blood	13.0	3.0	—	More soluble than blood meal.
Fish scrap	7.75	13.0	3.80	Use in compost or turn under soil several months before planting; usually slightly alkaline.
Greensand	—	1.5	5.0	Mined from old ocean deposits; used as soil conditioner; it holds water and is high in iron, magnesium, and silica.
Chicken manure[c]	1.65	1.5	.85	Dried; fast-acting fertilizer. Breaks down fastest of all manures.
Horse manure[c]	1.65	1.5	.85	Fresh, 60 percent water; medium breakdown time.
Oak leaves	.80	.35	.15	Break down slowly.

Table 22 (Cont.).

Some Common Organic Fertilizers

Fertilizer	Percentage[a] of			Remarks
	N	P[b]	K[b]	
Pig manure[c]	.5	.3	.5	Fresh, 85 percent water; balanced fertilizer; medium breakdown time.
Rabbit manure[c]	2.4	1.4	.6	Fresh; most concentrated of all farm manures.
Crustacean shells	4.60	3.52	—	Shrimp, lobster shells; break down slowly; contain large amounts of lime.
Wood ashes	—	1.50	7.00	Very fast-acting; alkaline; contains many trace elements.
Night soil[c]	.51	.10	.23	Human dung, fresh, 80 percent water; should be well-composted or pasteurized to prevent disease; breaks down slow to medium.
Feathers	15.0	?	?	Break down slowly unless washed of oils.
Hair	17.0	?	?	Cheap fertilizer and soil conditioner; oils slow breakdown.
Granite dust	—	—	5.0	Gradually available.
Rock phosphate	—	33.0	—	Gradually available.

[a]These are average percentages; the amount of nutrients in a particular sample may vary.
[b]This table lists phosphoric acid (P_2O_5) and potash (K_2O); to convert to their elemental states (P and K), divide the above figures by 2.3 or 1.2, respectively.
[c]The percentage of nutrients in manures depends on water content. Commercial manures are usually rated higher; for example, cow manure, 1.5-.85-1.75.

slowly and are available only after a period of time. Others are low or lacking in one or more of the major nutrients. Organic fertilizers can be combined to provide a complete balance.

Chemical Fertilizers

Most chemical fertilizers act quickly because all the nutrients are in soluble form. They are usually more concentrated than organic fertilizers, and can toxify the soil and kill the plants when they are overused. Fertilizers come in various concentrations and ratios of nutrients. All packaged fertilizers list the percentages of N-P-K (actually N-(P_2O_5)-(K_2O)). Also listed is the potential acidity or alkalinity, that is, the number of pounds of lime or sulfur required to counteract pH changes caused by the fertilizers. Chemical fertilizers are often incompatible with each other; so home gardeners who use them should buy them premixed or as a complete component fertilizer set.

Solubility is a major problem with commercial fertilizers. In irrigated areas as well as areas with rainfall during the growing season, they are likely to be leached away; so they must often be applied several times during the growing season. A typical program might be to fertilize at planting and every six weeks thereafter until the beginning of flowering. When spreading fertilizers during the growing season, do not let them come into direct contact with the roots. An easy way to fertilize during the growing season is to make a small trough between rows with the corner of a hoe. Fertilizer is placed in the depression. Some new chemical formulas release nutrients during the length of the growing season, and therefore need only one application.

Amounts to Use

The amounts of nutrient needed per acre and per 1,000 sq.ft. are shown in Table 23. Soils rich in one nutrient may be average or deficient in another. To calculate the required amount of a specific fertilizer, *divide* the amount of nutrient required as listed in the chart by the percentage of nutrient in the fertilizer. For instance, to add 5 lbs of N to an area by using bloodmeal, divide 5.00 by 0.15. The total comes to a little more than 33 lbs. Dried cow manure contains about 1.5 percent N. About 333 pounds of it are needed to supply 5 lbs of N. Urea, a chemical fertilizer, contains 46 percent N. Only about 11 pounds are required to supply 5 lbs of N.

Planning a Garden Fertilizer Program

Now let's plan some garden fertilization programs, to help some cultivators in three areas which have different soils and climates: New England, Kansas, and Florida. We'll see how growers with different goals adjust their garden soil.

Table 23.

Suggested Amounts of Nutrients
to be Added to Various Soil Types

Soil Type	Area[a]	N	P	K
Rich	acre	—	—	—
	1,000 sq.ft.	—	—	—
Medium	acre	50 lbs	10 lbs	40 lbs
	1,000 sq.ft.	19 oz.	4 oz.	1 lb
Fair	acre	100 lbs	15 lbs	80 lbs
	1,000 sq. ft.	37 oz.	6 oz.	2 lbs
Poor	acre	150 lbs	20 lbs	120 lbs
	1,000 sq.ft.	56 oz.	8 oz.	3 lbs

[a] 1,000 sq.ft. is, for example, a garden 20 by 50 feet. An acre is 43,560 sq.ft. or a square 208.7 by 208.7 feet.

New England   Most New England soils, and many soils in humid temperate areas, have a thick layer of humus which supplies N. New England soils also contain moderate amounts of P, but they are low in K.

Our first gardener has a typical New England soil in his backyard. From tests and observation he thinks his soil contains moderate amounts of N and P, but is low in K. A test indicated a *p*H of 5.8. He plans to start preparing his ten-foot-square plot (100 sq.ft.) in the fall, before frost. By planting time, he expects his backyard garden to have a *p*H of 6.7 and a balanced, fertile soil.

From Figure 59 he finds that the soil requires about 8.1 lbs of lime. He has decided to adjust the *p*H by using dolomitic limestone (with a calcium carbonate equivalent of .45) because farmers in the area sometimes complain of Mg deficiencies. Dividing 8.1 by .45, he finds that the soil requires 18 lbs of limestone. (Lime requirement divided by calcium carbonate equivalent equals the amount of limestone needed.)

He guesstimates that the N content of his soil rates between fair and medium, and figures the soil can use almost .2 lbs of N. He has decided to spread fresh manure from a nearby stable mixed with lime. In the spring he will turn this into the soil; at the same time, he will add manure composted with hay and

BOX F

Chemical fertilizers usually supply P in the form of superphosphate or triple superphosphate. These chemicals are manufactured by mixing rock phosphate with acids. Potassium is supplied by means of muriate of potassium (K and chlorine) or sulfate or potash, which are mined in the Southwest and purified. All these chemicals are soluble and are available to the plant. But a portion of them gradually reacts with the soil and becomes fixed or unavailable. As this portion becomes unavailable, it increases the total reserve in the soil, which reaches a new balance of available to unavailable nutrients. This results in a higher total of available nutrients than before fertilization.

Bone meal and rock phosphate, the most commonly used organic sources of P, and granite dust, a source of K, are not readily available, but increase the total reserve of nutrients and gradually increase the total amount of available nutrients. However, there is some time lag before these nutrients are available to the plant. They are usually applied in large amounts, at about three times the weight calculated for fertilizers of that concentration. But one treatment lasts four years or more, because the fertilizers remain fixed in the ground until they are used.

Table 24.

Percentage of Available N, P, and K in Several Organic Fertilizers

Fertilizer	N	(P_2O_5)	(K_2O)
Burned bone, ground	—	34.70	—
Fish scrap	7.76	13.00	3.80
Greensand	—	1.50	5.00
Molasses residue	0.70	—	5.32
Tankage	6.00	8.00	—
Wood ashes	—	1.50	7.00

table scraps. The fresh horse manure contains about .44 percent
N. To find out how much manure he needs, he divides .2 (the
amount of N required) by 0.0044. The total comes to about
45.5 lbs. (Nutrient required divided by percentage in fertilizer
equals amount of fertilizer needed.) The manure also contains
.17 percent phosphoric acid (P_2O_5) and .35 percent potash
(K_2O), referred to hereafter in this chapter as P and K, respectively. Multiplying .17 percent (.0017) and .35 percent (.0035)
by 66 lbs, he finds that he has added .11 lbs of P and .23 lbs of
K. (Lbs of fertilizer times percentage of nutrient in fertilizer
equals amount of nutrient in fertilizer.)

From Table 23 he finds that the soil requires about five
ounces of P. How many ounces of P is .11 lbs? He multiplies
.11 by 16, the number of ounces in a pound, and finds that the
total is about 1.75 ounces. The soil requires another 3.25 ounces.
Bone meal is about 20 percent P. To supply three ounces of P,
about a pound of bone meal is required. But bone meal breaks
down slowly, and is therefore applied at three times the rate
used for other fertilizers; so our cultivator uses 3 lbs.

Since the K content of this New England soil is poor, about
.3 lbs of K is required. The manure has already supplied .2 lbs;
so the soil requires another .1 lb. Our cultivator decides to use
wood ashes from his fireplace. Wood ashes are about 7.0 percent
K. He divides .1 by 7 percent (.07) and finds that the soil can
use at least 1.4 lbs of ashes. He adds this in the spring just before planting, because the ashes are highly soluble. Over the
winter, such highly soluble nutrients would leach away or become unavailable.

Our grower knows that some of the N in the fresh manure
that was added in the fall will leach away during the winter. But
the manure compost that he adds in the spring will more than
make up for any losses.

A New England farmer not far from the cultivator has been
rotating his fields from corn and marijuana to alfalfa and pasture
for the past ten years. Each fall he adds 7 tons of manure per
acre. Except for occasional additions of lime, no other fertilization is necessary.

A rural New England grower has decided to plant in a remote mixed-forest area. The first 10 inches of soil is a rich
compost of humus. It is full of life: insects, worms, and other
creatures. The grower has decided to increase the fertility of the

soil by using chemical mixes and dolomitic lime. He is cultivating in three clearings with a total area of about 1,000 sq.ft. He guesstimates that the soil is *medium* in N and P, but *poor* in K. It is also acid. He applied enough lime to correct the soil's natural acidity and the *p*H of the fertilizer.

Using Table 23, he decides that he should purchase a mix with a ratio of 50 parts of N, 10 parts of P (reading these from the *medium* line), and 120 parts of K (from the *poor* line), that is, a ratio of 5-1-12. A local nursery sells a commercial fertilizer with nutrient percentages of 10-5-25, close enough to the desired ratio. By taking the total amount of N required for a medium soil as listed in Table 23 (19 ounces), and dividing it by the N in the fertilizer (10 percent or .10), the rural grower finds the total amount of fertilizer required (190 ounces, or a little less than 12 lbs). The other nutrients are automatically added in the same ratio.

Kansas A cultivator in Kansas decides to plant along a hidden stream bank. The banks are covered with lush vegetation as a result of runoff that contains soluble fertilizers used on nearby farms. The cultivator feels that additional fertilizers are not necessary, since the vegetation is so lush.

Another grower in Kansas found that her soil was very low in N and P, but high in K, typical of dry midwestern and western soils that support scrub vegetation. It had a nearly ideal *p*H. She started to prepare her 200 sq.ft. garden in the spring after the rain season ended. Using Table 23, she found that it required 3.5 lbs of N, 6 ounces of P, and no K. Activated sludge (5-3-0) was available at the local garden center. To find out how much sludge her garden required, divide 3.5 by 5 percent (or .05). The total comes to 70 lbs.

Florida A grower planting 500 sq.ft. on a deserted ranch in central Florida started with a very sandy soil whose *p*H was 4.9 because of sulphurous water in the ground. From Figure 59, she found that the soil required about 35 lbs of lime. To adjust the *p*H, she used 14.0 lbs of a limestone with a calcium carbonate equivalent of 2.5.

The soil had virtually no organic matter, and she was not sure she could use the same location next year; so she decided to apply soluble mixes throughout the growing season. From Table 23, she found that "poor" required 28 ounces of N, 4 ounces of P, and 24 ounces of K. A chemical fertilizer with nutrient percentages of 15-5-10 was on sale at a local discount store. To find out how much fertilizer is needed to supply 28 ounces of N, divide 28 by 15 percent (or .15); the result is about 186 ounces of N, or about 11.5 pounds. Since the other nutrients are supplied at the same proportions or at higher proportions than are required, no supplements are needed at planting time. But additional feedings will be required periodically during the growing season.

TECHNIQUES FOR PREPARING SOILS

Each garden situation is unique, and many factors help determine which garden techniques you should use. These include the soil's condition, the size and location of the garden, commitment, and personal preferences. Each technique affects the microecology in its own way. Home gardeners may use techniques that are impractical for a farmer or guerilla planter. But all growers have the same goal when they prepare soil for planting: to create a soil environment conducive to growing a healthy, vigorous plant.

BOX G

Fertilizing *Cannabis* Depends on the Crop

Historically, *Cannabis* is known to require high fertility. In a fertile soil, *Cannabis* can outgrow practically any annual plant. *Cannabis* also is a known depleter of soils. This is true particularly with marijuana, since seeds, flowers, and leaves comprise the harvest. Hence it's necessary to fertilize the plants each year.

Hemp, on the other hand, comes from the *Cannabis* stem, and the fiber consists primarily of cellulose $(C_6H_{10}O_5)_n$. When hemp is grown, all plant parts except the fiber are returned to the soil; so the nutrients are also returned. Moderate fertilization, if any, is all that's required for hemp farmers.

If you are already growing a vegetable garden, the chances are that your soil is in pretty good shape for growing marijuana. However, vegetable gardens may be a little acidic, particularly east of the 100th meridian. The soil should be prepared in much the same way that it is prepared for corn cultivation, with the addition of lime to raise the pH to near neutral.

Tilling

Gardens which may not have been planted recently (in the last three or four years) require more work. It is best to begin preparing the soil in the fall, before the first frost. This can be done using a spade or shovel. The ground is lifted from a depth of six or eight inches and turned over so that the top level, with its grass and weeds, becomes the bottom layer. Large clumps are broken up with a blade or hoe. Larger areas can be turned with a power hoer or rototiller. Conditioners, such as fresh leaves, composts, mulching materials, pH adjusters, and slow-release fertilizers are added and worked into the soil, so that they begin to decompose during the winter. It is especially important to add these materials if the soil is packed, mucky, or clay-like. Soluble fertilizers should not be added in the fall, since they leach to the subsoil with heavy rains.

In the spring, as soon as the ground is workable, turn it once again. If the soil still feels packed, add more conditioners. If you are using manure or other organic materials, make sure that they are well decomposed and smell clean and earthy. Fresh materials tie up the N in the soil while they cure, making this nutrient unavailable to the plants. Commercial fertilizers and readily soluble organics, such as blood meal and wood ash, are added at this time.

The ground can also be seeded with clover or other legumes. Legumes (alfalfa, clover, vetch, etc.) are plants which form little nodules along their roots. The nodules contain bacteria which live in a symbiotic relationship with the plant. As part of their life processes, these bacteria absorb gaseous nitrogen from the air and convert it into a chemical form the plant can use. During its life cycle, clover uses up most of the N, although some leaks into the surrounding soil. But when the plant, or any of its leaves, die, the contents become part of the soil. The process of growing a cover crop and turning it into the soil is sometimes called green manuring.

After the last threat of frost, at about the same time that corn is planted, the soil should be worked into rows or mounds, or be hoed. At this time, the seeds should be planted. If any concentrated fertilizer is added to the soil, it should be worked into the soil and should not come into direct contact with the seeds.

The actual amount of tilling that a given soil requires depends on soil condition. Sandy soils and light loams may need no turning, since they are already loose enough to permit the roots to penetrate. Turning may break up the soil structure, damaging its ecology. These soils are easily fertilized, by using soluble mixes or by the layering technique described below. Soils which are moderately sandy can be adjusted by "breaking" them with a pitchfork; the tongs are pushed into the ground and levered or pushed, but the soil is not raised. This is done about every six inches and can be accomplished quickly. Farmers can loosen sandy soil by disking at five or six inches.

Some gardeners mulch the soil with a layer of leaves or other materials to protect it from winter winds and weather. This helps keep the soil warm so that it can be worked earlier in the spring. In states that border west of the 100th meridian, this helps prevent soil loss due to erosion from dry winds. Soil often drains well in these areas, and the ecology of the soil is better served when it is left unturned. At season's end, marijuana's stem base and root system are left in the ground to help hold topsoil. The next year's crop is planted between the old plant stems. Other gardeners plant a cover crop, such as clover or alfalfa, which holds the soil and also enriches the nitrogen supply.

Layering

Layering is another method of cultivation. The theory behind this program is that in nature the soil is rarely turned, but builds up, as layer after layer of compostable material falls to the ground. This material, which contains many nutrients, gradually breaks down, creating a rich humus layer over a period of years.

The layering method speeds up the natural process. Since gardens are more intensely cultivated than wild fields, new material is required to replenish the soil nutrients. Gardeners like Ruth Stout "sheet compost," that is, they lay down layers of uncomposted material and let it decompose at the same time that it serves as a mulch. But most gardeners prefer to use material which is already composted. The compost shrinks and builds the topsoil layer about an inch for every six inches of compost. After several years, the soil level will be raised considerably, and the top layers will be an extremely rich, porous medium which never needs turning. In order to prevent a spillover of the soil, gardeners usually construct simple beds (using boards) to contain the garden areas.

Layering is most successfully used on porous soils, especially sands, which contain little organic matter. It can also be used with clay soils. However, experienced growers say that clays should be turned several times before the technique is used, or the first couple of harvests will be small.

Planting a cover crop such as clover will give the soil structure. As more compost is added, the clover is covered and the new seed planted. The clover, with its N-fixing properties, remains a permanent cover crop. When marijuana seeds are to be planted, a planting row is easily tilled with a hoe. The clover protects the soil from sun-baking and its resulting water loss, and makes it harder for weed seeds to get started.

Tilling and layering are basic methods which are used with many variations. In some ways, there almost seem to be as many gardening techniques as there are gardeners. For instance, one gardener bought three cubic yards of topsoil and a cubic yard of composted steer manure. He mixed the material and filled raised beds with it to a depth of 18 inches, and had an instant high-power garden. Another grower made compost piles in his raised troughs during the winter. By planting time, the compost was complete and filled with earthworms. The beds became warmer earlier, and he could plant sooner.

A midwestern gardener used marijuana as a companion crop in much the same way Indians used corn. In between the marijuana, she planted beans and squash. She didn't get many stringbeans and only a few squash. But she believes that the beans gave the plants extra N, especially during the first six weeks, and the broad squash leaves protected the soil from the hot August sun.

A gardener in Georgia had such a sticky clay soil that a shovel once got stuck in it. He dug holes two feet deep and two feet wide with a power auger and filled them with a fertile mix of two parts sand, one part clay, three parts topsoil, and one part chicken manure. He claimed that his plants grew six feet in 10 weeks. Filling holes with a rich soil mixture is popular with guerilla farmers, who often must plant in poor native soils.

Mulching

Mulching is a labor-saving technique that many gardeners and farmers use for a multitude of reasons. A mulch placed on the ground before fall frosts helps the soil retain heat and protects it from winds and freezing temperatures. In the spring the mulched soil becomes warmer earlier in the season, and can be planted several weeks sooner than usual. A mulch cover keeps the seedlings' roots warm and eliminates a lot of weeding, since most weed seedlings cannot pierce the cover.

During the summer, mulches keep the ground cooler and more moist by absorbing and reflecting light and reducing surface evaporation. These are important points for farmers in dry areas. The water savings can be 50 percent or more.

Any plant or animal material will do for mulch. Gardeners use hay or straw, leaves, composts, manures, sawdust, bark, or plant clippings in two- to six-inch layers. A barber in Palo Alto uses hair. Baled hay is inexpensive and easy to use as a mulch. Round hay bales unroll in a long sheet that is easy to spread over the ground, and square bales can be pulled apart into tile-like squares.

Mulches create an ideal environment for earthworms and microorganisms which condition and enrich the soil. These organisms require a relatively cool, moist, dark environment. The mulch develops a dry outer crust which reflects light, keeping the underlayers cool and moist. Materials such as leaves, bark, and sawdust decay slowly because they do not contain enough nitrogen to maintain dense populations of decomposing microorganisms. Manures and composts contain more nitrogen and decay more quickly.

With few exceptions, mulches can be applied practically any time of the year, but the best time is probably in the fall, after the crop is harvested and before the ground has frozen. Leaves, plant clippings, and straw are applied in a thick layer from six to ten inches deep. Hay is layered two to six inches deep. Denser substances, such as manures and composts, should be mixed with straw and leaves to aid decomposition. This mixture is spread in an even layer, about two to four inches deep, over the entire surface of the garden. If winds pose a problem by blowing the mulch away, you can cover it with newspapers or sheets of plastic held down with rocks. If your area is dry, give the mulch a good soaking once before frosts.

By the spring, much of the material will seem to have disappeared. But underneath the top layer, you will find a soft-textured, earthy-smelling humus, teeming with worms, insects, and other small animals. This is a sign of a healthy ecosystem and a fertile soil.

Some people apply mulch in the spring, placing it between rows as they sow the seeds. The mulch keeps weeds from competing with the seedlings, absorbs the sun's warmth, and releases nutrients to the soil.

In cold areas, such as Montana, New England, and Alaska, growers place black plastic sheets over the soil. These absorb the sun's heat, allowing the soil to be planted sooner. The seedlings develop quickly in the warmer soil. The plastic is removed once the seedlings are well-established.

Newspapers and white plastic can be used to decrease water loss during the summer. They also reflect light back to the plants.

One innovative grower from western Colorado placed a sheet of white plastic over her garden and cut out holes wherever she planted the seeds. Though it is quite dry where she lives, she didn't need to water the plants until late July. And she had no problems with any weeds.

Containers

Containers are another option open to growers. Plants can be grown full-size in containers which are at least five gallons (larger would be better). Fill them with high-grade topsoil, or a planting mixture as described in Chapter 6. Planters are a convenient compromise where the soil is particularly poor or for the home gardener who does not wish to get into large-scale gardening. But remember, eight good-sized plants can yield over four pounds of grass.

Plants in pots need to be watered frequently, but require much less total water than a garden. The gardener can also move the plants. Some gardeners use this technique to maximize the amount of sun the plants get during the day, or as the sun's position changes with the season. And growers can easily induce early flowering by moving the plants to a darkened area.

Figure 61. Containers are convenient for outdoor gardens.

Almost any large container that can withstand the weight
of moist soil and which has holes for drainage is suitable. Containers which held toxic chemicals, herbicides, insecticides, or
other possibly harmful substances should be avoided.

We have seen all kinds of ingeniously made containers. Some
growers use old bathtubs, and others use wooden packing crates
or bushel baskets. A simple wood container 18 inches wide,
eight feet long, and 18 inches deep was made by a New Jersey
grower, who grew six plants in it. Trash cans, plastic containers,
barrels, and even rubber tires have been used. One grower grew
plants in one-cubic-foot bags of soil by cutting a five-inch-dia-
meter hole in the top and poking holes for drainage. To assure
drainage, growers sometimes fill the bottom of each container
with a six-inch layer of stones or gravel; if you are planning to
move such containers, lightweight perlite would be more suitable.

GUERILLA FARMING

Guerilla growers often use the same techniques as home gardeners. But the soil that they start with is sometimes marginal,

and the gardens are in remote, hard-to-get-to areas; so they modify the techniques to fit their needs. When it is impractical to carry bulky organic fertilizers to the growing site, guerilla farmers use highly concentrated commercial mixes. Compost and soil adjusters are gathered from the surrounding area, and the simplest, most light-weight tools are used. Some growers use horses or mules to carry equipment and material, and then use the animal to plow. The animals are quiet and, naturally, require no external power source. Experienced growers say that the animals can work as fast as or faster than a rototiller.

It is hard to generalize about details of guerilla farming, since much depends on the specific circumstances, which can vary greatly. For instance, a grower who plants along the fertile bank of a midwestern stream may not need to do more than pull out weeds and till the actual planting area. But a grower planting on a mountain slope may have to "build a soil," since soil and nutrients are washed from the slopes and down to the valleys by rainfall. For this reason, we will cover several situations separately: forest; washed-out steep areas; swamps and marshes; stream banks; grasslands and fields; and arid soils.

Forest Soils

Clearings in forests have always been popular places to plant because they offer security from detection. They vary greatly in drainage qualities, fertility, and pH. The drainage qualities of forest soils depend on the depth of the humus layer and the structure of the underlying subsoil. But most of the forest remaining in the U.S. is sloped, and water that is not absorbed by the soil runs off.

Soils are created in forests from the leaves, branches, animal droppings, etc., which accumulate on the forest floor. The first trees to grow are long-leaf pines, such as jack pines, which can grow in relatively infertile soils. Their roots penetrate deep into the subsoil to obtain some nutrients. Short-leaf pines, conifers, and firs appear as the humus accumulates, since they require a more fertile soil than long-leaf pines.

Pine-forest soils vary in fertility from poor to fair, and are usually quite acidic. In the Northeast their pH may be as low as 3.5, but generally the pH ranges from 5.0 to 6.0. In order to support a high-energy, lime-loving crop like marijuana, they require fertilization and liming. Long-leaf pines sometimes grow in compacted clay soils, which also require tilling.

As the soil evolves, deciduous trees (trees that drop their leaves each winter), such as oak and maple, may begin to grow. Deciduous forests, sometimes called broad-leaf or hardwood forests, have the best soils. These forest floors are covered with bushes, grasses, mosses, and other small plants. They have an adequate rainfall and a humus-rich soil, which is porous, holds water well, and can support a healthy marijuana crop, although additions of nitrogen fertilizers would probably spur growth. Hardwood forest soils have a pH range from 6.0 to 7.5. The soil in timbered forest land has a much smaller humus content, especially if it has been clearcut.

Mountain Soils and Washed-Out Steep Areas

Mountain slopes characteristically have little soil matter; their surface is composed largely of rocks, gravel, and sand. For long-term use they could be terraced so that newly formed soil is not washed away, but most growers are interested in more immediate results. These "soils" do not provide much of an anchor for marijuana's taproot and do not permit a network of lateral roots to form. Many of these soils also suffer from a low water table, since they drain rapidly. But there may be some sand and a bit of organic matter built up along gullies or in depressions or other natural traps. Such soil has usually had most of its nutrients leached out, but may contain some phosphates and potassium and considerable amounts of trace elements. The easiest way to adjust these soils is to use a well-balanced, slow-release, concentrated fertilizer. Bloodmeal, with its high N, works well with these soils.

One grower in the badlands of North Dakota used a timed-release 32-19-26 fertilizer in his "rock garden." He spread it just below the surface at the beginning of the growing season. Every time that it rained, his plants received nutrient-rich water. Toward the middle of the season, he noticed the lower leaves begin to pale, so he fertilized them periodically with urea. Heavy rains leach soluble fertilizers away, and in rainy areas they need to be applied three or four times during growth.

Containers can also be used in this environment. Growers use plastic bags or folded milk cartons instead of backpacking with a column of containers. When they get to the site, they fill the bags with a mixture of sand, as much as they can find, and gravel. The greater the ratio of sand to gravel, the longer the container will hold water.

One grower doublelayers heavy-duty polyethylene bags, and lines them with heavy-duty paper cement sacks or burlap bags. He fills the bag with gravel, then pours in sand and shakes it. He says that the mix is just about right when it looks like a can filled with gravel with sand in the spaces. He carries on a watering and feeding program much as he would for any hydroponic system.

Swamps, Marshes, or Bogs

These soils are very high in fibrous organic material, but are low in calcium and in available N, P, K, and Mg, which are leached from the soil or are insoluble because of the low soil pH. Since these soils are constantly wet, *Cannabis* roots cannot come in contact with air; as a result, the plant's growth is stunted, and the lower stem becomes susceptible to stem rot. These soils need to be adjusted to support a healthy crop of marijuana; they must be drained, fertilized, and limed. On a small scale, the easiest way to modify them is by constructing raised mounds, hills, or rows, at least one foot wide at the top and two feet high. The raised areas drain well, leaving relatively dry soil. Wood chips, chopped brush, sawdust, or perlite may be added to keep the mound light and the soil loose and aerated.

Wet soils are usually highly acid and should be limed. Once the lime interacts with the soil, nutrients which were locked up become available to the plants. Since these soils are rich in organic matter and have a high rate of microbial action after they are loosened and limed, they may need little fertilization.

Grasslands and Fields

These soils are usually fairly fertile and can support a worthwhile crop with little effort. They are usually well-drained, although they may be a little too dry or too wet. (If they have unusually large numbers of earthworms, they are probably a little too wet.) Their pH is usually between 5.5 and 6.5, although it may range up to 7.0. These soils are usually loams, which need only tilling in a two-foot radius, three or four inches deep, around each plant. All weeds and grass should be pulled from the area. Some growers mulch the cultivated area with newspapers, leaves, or dead grass. A grower in the Midwest adds crushed eggshells and a commercial timed-release fertilizer when he plants. He feels that this "extra boost" makes the difference between an adequate crop and a bountiful crop. Other growers periodically fertilize with soluble mixes. Some of these soils have to be irrigated during the long summer droughts. If they aren't, the plants won't die, but they will not grow to full size.

Stream Banks and Canal Ditches

These are some of the most convenient areas for growers to plant, since they provide an ample supply of water, which may contain fertilizer runoff. Stream banks are an area that marijuana naturally colonizes, and the planter usually needs only to cultivate the area to be sown, and cut surrounding bush so that the young plants can compete with established plants. If the surrounding vegetation looks pale and stunted rather than lush green and vigorous, the soil should be fertilized. These soils are sometimes low in calcium, which dissolves readily in water. Lime should be added to correct for acidity.

Sometimes the ground is a little too wet early in the growing season, although it dries out later on. Planting on hills or mounds is often used to solve this problem.

Arid Areas

Soils which have a low water table and dry out by June or July need to be irrigated to grow marijuana successfully. When irrigation is not feasible, growers plant along drainage ditches, streams, and canals, or look for green spots which indicate springs or underground reservoirs. Other growers use containers to minimize water loss. One grower in Arizona dug holes two feet wide and three feet deep, and lined the sides with thin polyethylene. He said that when he watered during the summer drought, he did not lose much water to the surrounding soil.

Arid soils usually have little organic matter, and drain quickly with extensive runoff. Some of them have a subsurface layer of clay, and therefore hold water on the surface until it evaporates. In any case their texture can be improved greatly by working in organic matter. The soil should be loosened at least two feet down. This loosening allows the taproot to develop deeply so that it can reach underground water during the drought.

Arid soils more often drain well, are alkaline, and contain P, K, and trace elements, but are low in N. Fish meal, cottonseed meal, blood meal, or manure may be the only additive the soil needs.

Chapter Fourteen
Planting and Transplanting

After the soil is adjusted, you are ready to prepare it for planting the seed. Sowing is an important process, since the post-germination or seedling stage is the most critical one for *Cannabis*. You can increase the seedlings' chance of survival by sowing the seeds properly.

WHEN TO PLANT

Most hemp-growing manuals advise that the seeds should be planted about two weeks after the last threat of frost, which is about the same time that corn is planted. As a rule of thumb, you need not plant until this time in areas that have a growing season of five months or more. These areas include most of the United States, except for Zone One (see Figure 62) and mountainous areas of the country.

Growers in northern areas report that plants have survived light evening frosts with little or no damage. We think of marijuana as a tropical plant, experiencing no chills in its native climes. But the mountainous areas of marijuana cultivation in Mexico and Colombia often have frosts during the growing season. One grower, describing spring (April) conditions in Nebraska, reported "plants (from *tropical* seed) three and four inches tall were covered with snow in the evening. By midafternoon all the snow had melted, and those little sprouts were healthy as could be."

Early-season sprouts do face more risks than later-germinating plants do. A lingering freeze or chill can weaken or kill them. Sometimes seeds or seedlings get washed away by heavy rains or flooding, or become infected from wet soil. They are also prey to hungry herbivores, who savor the tender young shoots, especially in the early spring, before the native plants have sprouted. These predators include rabbits, groundhogs, rats,

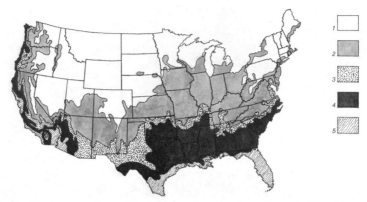

Figure 62. Average date of last expected spring frost. Zone 1, May 10 to June; zone 2, April 10 to May 10; zone 3, March 20 to April 10; zone 4, February 28 to March 10; zone 5, January 30 to February 28. Adapted from *Organic Gardening and Farming* magazine, April 1976. *Map by E.N. Laincz.*

mice, and possibly squirrels and cats, as well as large animals, such as deer, cattle, and sheep. Birds frequently eat the seeds and young shoots, especially if the ground looks planted. Snails and insects, such as cutworms and leafhoppers, also eat seedlings. Don't let this impressive list of dangers dim your enthusiasm. Although these problems do occur, they can be controlled or prevented with a little bit of planning (see Chapter 16).

As you can see in the Spring Thaw Map (Figure 62), the last date of expected frost varies from early February in parts of Florida, Louisiana, Texas, California, and Nevada to mid-June in the coldest regions of New England and the Midwest. Planting time varies locally, as well as regionally. Fields which receive direct sun warm faster than partially shaded ones. Fields covered with a layer of compost or fresh manure, or with black plastic sheets, retain more heat and are ready to plant sooner than other fields. Mountainous areas often vary considerably in planting time. Higher ground usually stays cold longer than low-lying areas. Since soil is dark, it heats quickly when exposed to sunlight. Soil is usually warmer in the late afternoon.

The time that the soil warms also depends on the weather. During severe winters, a deeper layer of soil becomes frozen than during mild winters; so it will take longer to thaw. Soil below this layer is insulated by the ice and remains unfrozen. Spring weather, rainfall, flooding, and cloud cover also affect the soil's temperature.

Actually, the only way to know whether or not a field is ready to plant is to feel it and look at it. Examine the soil in early morning. It should be easy to work, rather than hardened from ice. There should be no large frozen clods of soil or other organic matter. There should also be no fine crystalline ice particles which glimmer in sunlight.

For fall harvest, sow outdoors after March 21, the first day of spring and the turn of the Equinox, when there are equal lengths of sun-up and sun-down. There are an additional 20 to 30 minutes of light before dawn and after sunset, for a total of 13 hours of daylight. When plants are started earlier, they may flower prematurely because of the short days. The plants may also be subject to sex reversal, and more males may develop.

There is little advantage to starting *Cannabis* before April. Each plant has a certain genetically defined potential for growth and size. As long as the plants have enough time to grow and develop, usually five or six months, this potential is realized (some Colombian and Asian varieties may need longer to develop). Plants started before spring grow no larger in size than plants started during April. The younger plants are virtually indistinguishable from the older ones by harvest, and plants which are started earlier face more risks of detection and destruction.

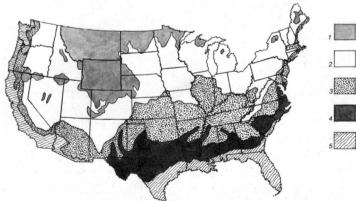

Figure 63. Average date of earliest expected fall frosts. Zone 1, August 30 to September 20; zone 2, September 30 to October 10; zone 3, October 20 to 30; zone 4, October 30 to November 10; zone 5, November 20 to December 20. Adapted from *Organic Gardening and Farming* magazine, November 1976. *Map by E.N. Laincz.*

However, if you are faced with a short growing season, you can get a head start by germinating the seeds a week to six weeks before the local planting time, and transplanting the seedlings outdoors at about the same time seeds would be planted in your area. You can also hasten planting time by covering the area to be sown or planted with a clear (or black) plastic sheet, which will warm the ground by the greenhouse effect.

PREPARING TO SOW

Growers use three basic techniques to sow marijuana: rows, hills, and broadcast. Each method is suitable within a certain range of conditions and has its own advantages and disadvantages.

Rows

Rows are convenient to use, especially for large areas. They are constructed easily using a hoe, plow, or tiller.

Rows facilitate the care of gardens and fields by setting up an organized space in which the plants and surrounding area can be reached easily by the gardener. Weeding, watering, thinning, pruning, and harvesting can be accomplished very quickly. Larger fields are planted in rows to accommodate plows, planters, and cultivators. They are essential when fields are flood-irrigated. Furthermore, they provide a way to use space in the most efficient possible manner. But rows make detection easier, since they have an orderliness that plants do not exhibit in nature.

On sloping and hilly ground rows are a major factor in soil conservation: such soil is easily carried away in windstorms and in the runoff after rain. For this reason, rows on hilly and sloping ground are contoured: curved to run perpendicular to the slope.

Space rows two to six feet apart; plant seeds every four to eight inches.* (See Box H.)

To construct a row, break up any large clods on the surface of the soil. In a garden-size area this is easily done by striking them with the tongs of a rake. In larger areas a tiller or externally powered cultivator can be used. Then level the soil.

If you need to irrigate or have problems with excessive mois-

---

*In any description of planting which we give, we refer to 100 percent viable seeds. In this case, for example, if seeds are tested (see Chapter 3) and have a viability of 50 percent, sow the seeds two to four inches apart. If they have a viability of 33 percent, sow them one to two inches apart.

ture, use a hoe to raise the soil into alternate rows of hills and trenches. Pat the crests of the hills with the hoe or a shovel so that they are an inch or wider at the top, and four to eight inches higher than the trenches.

BOX H

Plant Size and Spacing

Plants vary tremendously in size and branching habits because of many factors, including variety, soil fertility, length of growing season, amount of light received by the plant, water, spacing, and pruning. As a result, one can have no firm rule about how far apart plants should be spaced.

An individual full-grown plant may have a diameter at its base as wide as ten feet or as small as 18 inches. Most conical-shaped varieties (Colombian and Jamaican) grow between seven and 12 feet tall, and have a width between four and six feet. Mexican plants are somewhat taller and thinner, with a base diameter of three to five feet. Some exotic Indian, Central Asian, and Central African plants may have a diameter only one or two feet across. The descriptions are generalizations; there are many varieties within each country, and much variation within each variety.

Pruned plants have a much wider base than unpruned ones. Plants pruned at the fourth internode and again a month later sometimes grow twice as wide as an unpruned plant.

In order to catch as much sun as possible, rows should be oriented along a north-south axis, perpendicular to the course of the sun. The advantage of such rows is more pronounced in southern than northern latitudes, but the solar-energy differential in north-south versus east-west rows is significant at all latitudes in the United States, and becomes more important on steep slopes. Another factor is the orientation of the garden as a whole. Plants sown in a square plot whose sides point northeast and southeast get about 10 percent more light than ones in a plot whose sides point due north and due east.

Hills

Hills and mounds are especially convenient for small plots. Low hills are often camouflaged to look like natural or wild stands, and are very useful in areas in which the land is too wet in the spring, because the hills drain above the ground level. They are easily adapted to meet unusual requirements. For example, a grower in New Mexico planted a donut-shaped hill eight feet in diameter and two feet thick, leaving a center hole four feet in diameter. He placed a portable plastic tub in the hole after punching pinholes around the edges. To water he just filled the tub. In the swampy Everglades, two industrious farmers constructed a giant hill-row three feet thick and three feet high. The hill had such good drainage that it kept the plant roots well-drained even during the rainy season.

Hills are usually constructed between two and five feet in diameter. Small hills are usually planted with 15 to 20 seeds, and large ones may be sown with as many as a hundred. The hills are spaced three to 10 feet apart, so that each group of plants gets a maximum amount of light. Hills can grow more than you would at first suspect. For instance, if you were to grow a hill three by three feet, you could harvest six to nine large plants. Their foliage would extend two and a half feet beyond the hill, for a total of about thirty square feet of foliage space.

Broadcast Seeding

Broadcast seeding is the fastest and easiest way to sow, but is not an efficient way to use seed. Seeds are simply tossed or shaken onto the prepared ground, at the rate of about forty per square foot, and are then usually pressed into the soil with a light roller or by foot. This method is most effective in moist soils. Many of the seeds never germinate or die immediately after germination. The faster growing ones naturally stunt the others by shading them. This method is often used by guerilla farmers who want the stands to look natural and who wish to plant large areas quickly. An experienced grower can sow several acres a day by hand using this method.

Seed Count

There are approximately 2,300 medium-sized seeds in an ounce, or about 85 per gram. An acre is about 43,000 sq.ft., or a square

208 feet on a side. To plant an acre in rows two feet apart with a seed every four inches requires about 90,000 seeds or 39 ounces (1,100 grams, or two pounds, seven ounces). At this rate, a ten-by-ten plot requires about 2.5 grams of seed.

A typical hill field has four-foot-wide hills spaced about seven feet apart. A typical hill and surrounding area accounts for approximately 100 square feet. There are approximately 430 hills in an acre. If each of these is planted with 100 seeds, the field requires about 43,000 seeds, which weigh about 18 ounces.

Broadcasting requires a lot more seed. At the rate of 40 seeds per square foot, a grower uses about 2.3 ounces in a ten-by-ten plot. An acre requires about 47 pounds, or 21 kilograms of seed.

HOW TO PLANT

Finally, after the soil is adjusted, and the rows or hills are built, it is time to actually plant the seeds and watch your garden begin to grow. If you are growing with clover as a cover or companion plant, dig it up to a depth of four inches and chop up the soil. Water the soil to the point that it feels almost wet. Drill a hole with a seed drill, stick, or pencil, then drop one seed into the hole, cover it gently, and pat the soil down again. Marijuana seeds are large enough to handle individually; so each one can be planted separately.

How deep one digs the holes depends on the kind of soil in which one is planting. Light woodsy or organic soils are planted 1/2 to 3/4 inch deep, so that the stem is held firmly in an upright position. Sands and light loams are planted 1/2 inch deep. Heavy loams and clay are planted 1/4 to 1/2 inch deep, so that the sprout's energy is not expended before it breaks through the soil.

If you are broadcast seeding, you can increase the germination rate tremendously by screening a layer of soil over the seeds to help keep them moist. Seeds that dry out weaken or die.

In a garden that has been mulched, lift away the mulch cover at each place you plant, and sow the seed in the underlying soil.

In soft-textured soils, instead of digging or poking holes, press each seed to the desired depth, and cover or pat the soil smooth.

GERMINATION

The seeds need constant moisture in order to germinate. Therefore, the ground should be well-watered. Keep the soil moist by watering it with a light spray whenever it begins to feel dry. This may mean watering the immediate area once a day. You can keep the soil moist and hasten germination by covering the planted area with transparent glass or plastic. Most of the seeds should sprout in a period ranging from three days to two weeks. This variation depends on variety, age and condition of seed, and soil temperature; the warmer the soil, the faster the rate of germination.

Once they have germinated, the seedlings should be kept moist until the roots grow deep enough to absorb an adequate supply of water from the subsoil. If the ground is still moist from spring rains, as it is in many of the eastern regions, you may not have to water at all. On the other hand, there are sections of the West which are completely dependent ôn irrigation.

When the seedlings are only an inch or two tall, you can protect them from heavy rains or frosts by using drinking glasses, jars, or paper or plastic cups. You can protect larger plants with containers from which the bottoms have been removed. Transparent containers warm the soil by the greenhouse effect, capturing light and turning it into heat. In warm weather, use white or translucent containers, which prevent burn by reflecting some light and diffusing the rest. Containers also keep the soil moist, serve as plant markers, and protect the plants from some enemies. A grower in Berkeley, California, used cracked fish tanks to protect plants in the early spring. A guerilla farmer in the Poconos puts up four posts, one at each end of a row. She uses them as a frame for clear polyethylene covering, creating a small greenhouse.

Growers in Zone Five sometimes harvest a spring crop by transplanting indoor-grown, two-month-old plants outdoors right after the last frost date. The naturally short days and long nights trigger the plants into flowering. (See Transplanting, below, and the discussion of the photoperiod in Chapter 3.)

If started after May 15, marijuana may not have time to reach its full size or flower. This problem mainly affects growers in Zone One and in mountainous areas. But even if the plants do not grow to full size or flower, you can still harvest a potent crop of preflowering tops, which may be almost as potent as ripe buds. The harvest is not as large as a crop of buds, but it is more than worth the effort.

TRANSPLANTING

Seedlings and young plants are transplanted after the last threat of frost. If the growing season in your area is less than five months, you may want to start the plants indoors, or in cold frames, transplanting when the weather permits. A 10-by-four-foot cold frame can easily hold 60 two-month-old plants. The cold frame can be constructed with two-by-two's or branches gathered at the site. Cover the frame with a double layer of six- or eight-mil polyethylene plastic or similar material. Attach the plastic to the frame with tacks or staple-gun tacks. If the area is unprotected from the elements, slant the roof so that rain will run off. If the area is windy, place rocks or branches along the frame to add weight. Orient the cold frame to face the south.

In areas with a growing season of six months or more, plants will not necessarily get larger if they are started earlier than normal. Plants started at normal planting time catch up to the older plants by season's end. It serves no purpose to start plants before about March 21, the spring equinox.

Where there is no threat of frost (in Hawaii, southern Florida, and parts of Texas, Louisiana, and California), growers can raise a winter crop. Grow the plants for two or three months under artificial light. Plants get off to a faster start under artificial lights than natural light during the winter months. Move or transplant them before the beginning of March. Most strains will flower because of the short days (less than 12 hours of light) and fill out to well-formed plants by the end of May when they are ripe.

For the normal summer crop, seedlings should be trans-

Figure 64. Simple-to-construct cold frame.

planted after the last threat of frost. The best time to transplant is on a rainy or cloudy day, which allows the plants to adjust to the new environment without the strain of intense sunlight. Plants grown in a cold frame or sunny window adjust more easily than plants grown under fluorescent lights. Plants grown under artificial light usually show evidence of shock when they are moved to sunlight. Near sea level they may lose some of their green color and appear pale or yellowed. At high altitudes, such as mile-high Denver, the leaves may actually burn, turn brown, and fall. Healthy plants usually recover quickly by adjusting the new growth to the changed conditions. However, plants can be conditioned to the new environment by being placed in a partially sunny area, preferably where they are shaded during the middle of the day and receive either morning or late-afternoon sunlight. The plants need about a week to adjust.

Seedlings grown in planting pellets for up to 10 days after germination can be placed directly in the soil. Peat pots should be scored with a knife so that the lateral roots can penetrate the pot more easily. Seedlings started in milk cartons or flower pots should be removed from the container so that the roots are disturbed as little as possible. Plan on using a pot size which is root-bound by the time that you transplant. (For the relationship between pot size and number of weeks, see Table 17.) To transplant, water the area to be transplanted and the plant.

Figure 65. Young plants raised with artificial light ready for transplanting outdoors.

Then dig a hole a bit larger than the pot and loosen the surrounding and underlying soil. Place the plant in the hole, and pack the soil so that the stem base is at the same depth that it was growing at before. Firm the soil and water the area.

In areas where ripoffs are expected, such as parts of Hawaii and California, some querilla farmers transplant individual plants (one to each site) to sites which are widely spaced over the countryside. In this way they may lose some, but at least not all, of their plants to ripoffs.

Each plant (one to three months old) is transplanted to a cone-shaped hole, two to three feet deep by two feet across the top. This strategy is well-suited to areas with poor soil. Since much of the hole is taken up by rootbound soil, it is easy to gather enough topsoil and sand to fill the hole. The gathered soil should also be mixed with organic or slow-release fertilizers which provide ample N and P.

Chapter Fifteen
Caring for the Growing Plants

WEEDING

Marijuana is a fast-growing annual whose survival depends on its ability to compete with other fast-growing weeds. At the end of each season, plants growing in a wild stand may cover the ground with thousands of seeds per square foot. Many of these are relocated by wind, runoff, and birds, and some are destroyed or die. Others never receive the conditions they need to germinate; and of those that do germinate, many die as seedlings. The remaining plants compete with each other and with other weeds for the available light, nutrients, and water. Even so, wild stands may be as dense as forty plants per square foot. In order to survive the competition, *Cannabis* expends a great deal of its energy during the first two months growing a main shoot which is taller than the surrounding vegetation. Then it develops lateral branches which shade the shorter plants. With their source of energy—light—cut off, the shaded plants stop growing and often die

When you cultivate—that is, eliminate weeds—the rate of germination and survival of your plants is increased enormously. Growers using clover, sheet composting, or mulch as ground cover can expect very little interference from weeds during seedling development. But plots of fertile, aerated, and cleared soil are open to colonization by a wide range of plants; so you may have to weed several times before the marijuana's dominance is assured.

When you weed, make sure not to pull out any weed seedlings which may have roots in the same area as the *Cannabis* roots. Instead, cut the weeds slightly below the surface with a clipper, scissors, or your fingernails. Weeds more than six inches away can be safely pulled. Leave them to dry right on the soil. As they dry and decay, they return the soil's nutrients to it.

Growers plagued with weeds can cover the soil with mulch, paper, or polyethylene sheets. One grower found that two computer sheets fit exactly between the rows. Another grower used torn drapes as a temporary ground cover.

Once *Cannabis* has established dominance over an area, the other weeds are not able to interfere with its growth. But if there is wide spacing between the plants, the weeds may have open space and start to grow rapidly. Keep these weeds clipped short if water or nutrients are scarce.

WATERING

Marijuana requires an ample supply of water to live and grow. The actual quantities that it needs depend on the plant's size, the gardening techniques, type of soil, temperature, wind, humidity, and intensity of light. A vigorous plant may transpire several gallons of water a day during the hot summer months. If it receives less water than it needs, it stops growing, wilts, and then dries out.

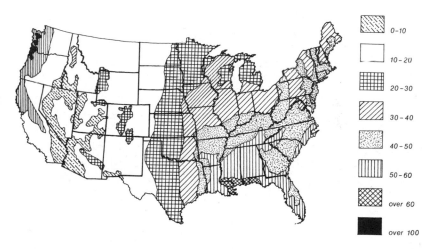

	0–10
	10–20
	20–30
	30–40
	40–50
	50–60
	over 60
	over 100

Figure 66. U.S. Rainfall Map. Areas with less than 30 inches of rain usually require some irrigation. Adapted from *Soils and Fertility*.[227] *Map by E.N. Laincz.*

Seedlings

Marijuana germinates best in a moist soil. Within a week, it grows a taproot three or four inches long. By the end of the first month, the root system may stretch over an area a foot and a half in diameter and go more than one foot down. Until then, the soil should not be allowed to dry out. Plants which have germinated during warm, sunny weather may need to be watered until the roots have grown deep enough to reach subsoil moisture. When the soil three inches below the surface feels dry, seedlings should be watered, preferably by using a watering can or the spray setting on a hose. Gently water the soil, making sure not to disturb the seedlings or the soil surrounding them. The soil should be thoroughly saturated so that the moisture percolates down, encouraging the roots to grow deep. If the surface is only lightly watered, the roots may grow near the surface, leading to water problems as the soil gets dryer during the summer.

After the first month, *Cannabis* does best when the soil goes through alternating moist and dry periods. This alternation allows the lateral roots to come into contact with air. By the end of the growing season, the root system may penetrate the soil to a depth of six feet or more. As long as they are not blocked by solid rock or dense clay, the roots grow by following a trail of moisture. If the trail leads deep, the roots follow. The deeper layers of soil are less likely to dry out during hot, dry weather; so plants which have a deep root system do not have to be watered as often as seedlings.

Older Plants

As a rule of thumb, *Cannabis* over a month old should be watered when the soil about six inches deep feels dry. But this rule provides only a rough indication that the plants need water, because there may be deeper sources of water that are not apparent. The most obvious indication of a problem is wilting. A more subtle one is slow growth during the (ordinarily fast-growing) vegetative stage.

Since you want to wet the lower layers, you should thoroughly saturate the soil. If the soil is completely saturated, it should hold water for a minimum of a week. Usually only two or three waterings a month are required by a garden that is completely dependent on irrigation.

The most efficient way to water is to let the water slowly seep into the soil, so that all the organic particles which hold the water are saturated. If the soil is very dry, and the water beads or runs off and is not absorbed, add household laundry detergent at the rate of one or two grams per gallon of water. It acts as a wetting agent, which breaks the surface tension. Once the soil is treated with a wetting agent, it usually absorbs water throughout the growing season.

In dryer areas where corn, cotton, and other deep-rooted crops are irrigated, marijuana also requires an additional source of water. But in areas where there are patches of wild hemp or where deep-rooted crops grow by using available ground water, marijuana does not need to be watered, although additional water may increase its growth.

Water conditions also vary from field to field. For instance,

---

BOX I

Water in General

As a general rule, areas east of the Mississippi River have a high water level and at least 30 inches of annual rain. As a result, there is water near the surface, and the plant roots do not have to grow very deep to reach it. After the first month the plants rarely need to be watered. Occasionally there is a dry summer, and the plants may need water then. When there is little rainfall, watering may increase the size of the plants, especially during the hot summer months, when the groundwater level drops and the soil is baked by the sun.

Areas west of the Mississippi typically have a much lower groundwater level than the East. Many of the population centers and agricultural lands in California, Arizona, New Mexico, and Utah are extremely dry, and support little vegetation without irrigation. The soils in these areas are frequently very porous and do not hold much water; so they must be watered often, sometimes even twice a week.

Deep soil layers retain water much longer than the top layers. To encourage the development of a deep root system, saturate the ground when you water. The roots follow the moisture trail.

many midwestern farmers plant along the banks of meandering streams. Even in dry areas, these plants have a natural source of water. Mountainous areas are usually well-drained and dry out before valleys do. Low-lying fields remain moist later, and are saturated by runoff from higher ground. In browned areas, farmers look for green spots which indicate underground streams, springs, or runoff. Planters look for deserted wells or active watermains with leaks. Fields high in organic matter retain moisture longer than other fields, and mulching may cut water evaporation by 50 percent.

Watering Techniques

Gardeners may supply water by using a bucket, can, or waterhose. But growers with larger plots often rely on waterpumps to deliver river, lake, or well water to their gardens. Irrigation canals, drainage pipes and ditches, and water mains are sometimes convenient sources of water. The two most efficient methods of watering are the drip hose, which seeps water around the plant, and hand watering into an enclosed area around the plant's stem.

There are several kinds of drip hoses. Some have perforations every three to six inches along their length. These are useful when marijuana is planted in rows or large hills. Another kind is actually a kit, consisting of a main feeder hose and several side hoses two to four feet long. Each side hose has a metal bulb at the end which can be adjusted to regulate water flow. The bulb lies near the plant stem. A drip bottle was invented by a grower in the dry area of Nebraska who was only growing a few plants. He punched pinholes in the bottom of several one-gallon milk jugs and placed a jug near each plant. The jugs slowly watered the garden. Every few days, he refilled the jugs from a nearby irrigation ditch. As the plants grew larger, he placed more jugs around them. The drip method moistens the soil slowly, but does not flood it; so the soil and its nutrients are not washed away. Since this method allows you to decide exactly where the water goes, you need not waste any on nonproductive land.

Growers sometimes use elaborate setups, such as battery-electric, hand- or foot-powered, ram- or windmill-driven pumps. Foot-powered pumps are probably the most convenient for

Figure 67. Containers with bottoms cut out conserve water.

small plots. They are extremely lightweight (just a little heavier than a bicycle), inexpensive, easy to construct and disassemble, and virtually silent. Since you have much more power in your legs than in your arms, foot-powered pumps can do more work, and do it faster, than hand-powered pumps.

Electric pumps are relatively quiet and pump an enormous amount for their small size. But they require a source of electricity. They cannot be used unless there is a power line available, although there are car alternators available which produce 110-volt current.

Gasoline pumps and electric generators are heavy and noisy. Even with a muffler, they can be heard for miles in some country areas. They require a source of fuel, and often an elaborate setup, including rigid feed tubing, fuel tank, and platform. But once they are in place, they can deliver a tremendous amount of water. They are usually used by farmers growing large plots. Sometimes growers dig a hole in which they store and run the equipment. This setup helps muffle the sound and keeps the machinery in good working order.

Ram- and windmill-powered pumps use running-water and wind energy, respectively. They come in many sizes and are often used to fill water tanks for later use. They can also be used to generate electricity to run electric pumps. They require no fuel, are usually silent, and can be constructed inexpensively.

Figure 68. Backpacking water pump. *Photo by A. Karger.*

Figure 69. Pump setup. *Photo by A. Karger.*

But some farmers have devised other methods for getting water to their plants.

A farmer growing near Tucson, Arizona, trucks water to her plants twice a week using a pickup truck and four 55-gallon barrels. She attaches a garden hose to her tanks, and siphons the water to her garden, 200 feet downhill.

Two foresighted farmers in Texas carried twenty 30-gallon plastic trash cans and lids to their garden. During the spring rains, they filled the containers from nearby gullies. By the end of the rainy season, they had collected enough water to carry them through the summer drought.

A homesteader in Oregon's dry eastern section dammed a gully by using an earth stabilizer, plastic, wood and cement, and pipe. During the winter his private reservoir filled.

Farmers near Atlanta tapped into a city water main. The pressure from the water main allowed them to pipe water uphill.

THINNING

If the soil is kept moist during germination, most of the viable seeds that you planted will germinate and the seedlings will soon start to crowd each other. This happens frequently when the plants grow on their own. Then they grow into a dense hedge-like mass dominated by a few plants. The dominant plants typically have long internodes and a long sturdy stem with little branching. The shorter, bushier plants are shaded by the taller ones and become stunted from the lack of light. By thinning, you give the plants that are left enough room to grow to their full potential, and you choose the ones that you think will grow to be the best for smoking. Leave the plants that have dense foliage, are branching, and, later in the season, the ones that are the most potent.

Thin the plants as soon as they begin to touch or crowd each other. This should be repeated as often as necessary. Seeds sown six inches apart in rows two feet wide require thinning several times during the season. But guerilla farmers sometimes let the plants compete so that the garden looks more like a wild stand.

There are two methods used to thin: cutting the stem at the base so that the entire plant is destroyed, and cutting just the tops so that the plant's growth is thwarted, and the uncut plants shade it. The cut plants remain relatively inactive, and do not use much water or nutrients, but they do shade the ground and use otherwise wasted space.

STAKING

Outdoor-grown plants rarely need staking. When the stem bends from the wind or rain, tiny tears in the structure develop. These are quickly mended by the plant: it grows new cells which increase the girth of the stem and make it stronger. But plants which are suffering from nutrient deficiencies or are top-heavy because of competition may need to be staked. Heavy rain sometimes causes the plants to fall over, especially if they have shallow root systems which cannot hold the added weight.

To stake, drive a sturdy rod six inches from the stem and deep enough into the ground to be able to give the plant support. Then tie the stem to the stake with wire twists or string.

If the stem or the branch is cracked, pinched, or bent at the base, its position should be corrected and held firmly with a splint. The splint can be held with masking tape. In a few days the plant grows tissue to support the damaged area.

PRUNING

Growers prune (clip or top) their plants to increase productivity, prevent detection, or to harvest early smoke. In the near future, new laws will decriminalize or legalize marijuana cultivation. These laws will probably limit legal cultivation either by the total gardening area or by the number of plants an individual or group may cultivate. Gardeners limited by space will maximize yield by cultivating a dense stand of tall, unclipped marijuana. Growers permitted to grow only a few plants will grow the largest, most productive plants possible. This is done by giving the plants the best possible growing conditions and a lot of space between plants to maximize light and minimize competition for water and nutrients.

Unpruned marijuana develops in one of three classic shapes, depending on variety. Many Mexican and Thai varieties develop into a tall, narrow bush no wider than three feet and shaped like a poplar tree. Colombian, Cambodian, Indian, and some south Mexican and Vietnamese varieties are Christmas-tree shaped. Some Moroccan and Afghani varieties have complex branching and naturally grow into small, dense bushes, about five feet tall. Marijuana usually grows to its full height by early September. Most of the marijuana plants you are likely to cultivate will grow to between eight and fifteen feet tall. Some Hawaiian and Thai varieties average between twelve and twenty feet tall.

Figure 70. *Left:* Early flowering pruned plant. *Right:* Not clipped.

Increasing Yield

When marijuana is clipped to increase the number of growing shoots, the total yield at season's end may not be increased. Provided that soil and water are not limiting to growth, each plant can reach a maximum size when given enough room. The more surface area the plant presents to light, the closer it will get to its maximum potential. Where the plants are grown with much space between them, clipped plants can yield more than unclipped plants, especially if the branches are spread out to maximize the light on the plant. When plants are grown close together, the taller a plant is, the more sunlight it will receive, and hence the larger the possible yield.

Some growers prefer to harvest a top stem that is thick with buds (colas). The largest colas form on the main growing shoot of unclipped plants. When the growing shoot is clipped from a plant, the new shoots and leaves grow slower and smaller than the main shoot of an unpruned plant because the capacity for growth is spread out over several shoots. When a plant is clipped early in the season, most of the difference in leaf and bud size is made up by harvest time.

Figure 71. Main stem is bent and side colas grow vertically.

Marijuana can be pruned at any time during the seedling or vegetative growth stage, but you should prune plants when they are young if you plan on harvesting growing shoots during the season. A seedling clipped anywhere from the fourth to the sixth node will usually form at least six strong growing shoots that can be harvested during the third or fourth month. If these shoots are cut again while the plant is still young, marijuana often develops into a small, very compact, hedge-like bush.

Yield can be increased by spreading the plant's branches so that more light reaches the inner growth. *Cannabis* stems are bent most easily when they are still green and fleshy, nearer to the new growth, but the whole plant can be bent to form a gentle arch with the top of the main stem in a horizontal position. Within a few days the side branches along the top will begin to grow vertically, competing with the main stem. They will soon develop their own horizontal side branches. To bend a plant, tie the main stem loosely with a cloth or heavy string. Tie the other end of string to a heavy weight or anchor on the ground. Don't put too much pressure on the stem as this tears some of the roots and weakens the plant. You can bend the plant a little each day until the plant is in the desired position.

You may also increase yield by bending only the growing tip. This encourages the side branches to develop sooner than they naturally would. Only the flexible part (about the last foot) is bent. To bend the top, use stiff wire or wire twists used for plastic bags and wrapping vegetables. Fasten the wire with a loop around the top of the shoot and bend the shoot gently so that the tip points down. Then fasten the other end of the wire lower on the stem to hold the tip in position. (See Figure 49.)

A common mistake that cultivators make is pulling off the large leaves on the main stem (sun or fan leaves), when the plants are young. These leaves are removed by cultivators who believe that their removal will cause the undeveloped side shoots to grow. But fan leaves are net producers of sugar and energy, which are used by the side shoots to begin growth. Rather than encouraging new growth, the removal of fan leaves slows growth. The plant will also be more susceptible to attacks from pests and predators.

When the plant is several weeks old and growing well, the difference between plants with their leaves removed and those left intact may not be large. The biggest difference can be seen when leaves are removed from branches just prior to, or during, flowering. The buds that form from leaf axils with leaves removed are noticeably smaller than those where the leaves have been left on the branch.

Detection

Cannabis can be detected from both the ground and the air. From the ground, marijuana is revealed by its familiar shape, unmistakeable leaves, and odor. Tall plants are usually more conspicuous than shorter ones. From the air, stands may have a different color than the surrounding vegetation, especially where natural vegetation is not as lush as marijuana. Individual plants usually have a circular profile when viewed from above; this can be altered by bending or pruning the plant. Varieties which are naturally tall-growing may need to be cut several times during the season to keep them hidden.

Plants are sometimes cut back severely, to as much as half their height when they get too tall, but this may damage the plant. A less drastic topping technique is to remove the top foot of growth. Whenever new shoots get too tall they are clipped. But plants should not be severely pruned late in the season when the growth rate has slowed (preflowering), because there will be fewer branches left on which buds can develop.

If you are trying to conceal plants behind a fence or wall, start bending or pruning the plants early, at about one month of age. By starting early and continuing to prune during the vegetative growth stage, you will train the plant to branch and fill up the area. If you wait until the plants are already tall, you may have to cut the plants back severely or clip shoots continuously.

GARDENING TIPS

Transplanting Older Plants

A friend of ours was warned that his garden had been spotted by local authorities. Rather than cut down his four-month-old plants, he decided to transplant them. He dug the plants out, leaving a ball of soil about two feet square around the roots of each one. He wrapped each soil ball tightly in a plastic bag to transport it, and placed the plants in newly dug holes in a different spot. He kept the plants well-watered. After a few days, they recovered from transplant shock and started to grow once again. Transplanting large plants is not easy to do, but it could save a crop. The marijuana root system is not very extensive when the plants are in fertile soil with plenty of water; the tap root may only be six inches long on a ten-foot plant.

Wind Protection

Hemp *Cannabis* planted closely together has been used by farmers to form a windbreak to protect other crops. If you are growing in an especially windy area such as the Midwest, you may wish to plant a perimeter of tightly spaced *Cannabis* to protect your garden. Construct a rope and stick fence against the windbreak to hold the plants upright and prevent them from falling into the central garden. Simply keeping the plants clipped short is a simpler approach.

Inducing Flowering

Growers may wish to induce their plants to flower early, especially in the North, where the growing season is short. Plants in containers can be moved to a dark area for 12 hours of darkness or more per day. Black sheets of polyethylene film, dark plastic bags, and large appliance cartons can be used to provide periods of uninterrupted darkness. Use the dark treatment nightly until the plants are flowering (usually after one to two weeks of long-night treatments).

Winter and Spring Crops

In southern parts of the U.S., Hawaii, and parts of California, you can grow more than one crop in a season. Greenhouses that stay above freezing can also be used for year-round growing. Plants started during the winter or early spring get naturally long nights and flower early, when they are relatively small, usually no more than four feet tall. Flowering can be postponed by breaking the long nights with short periods of light. This extends the vegetative growth period, yielding older, larger plants at flowering. Start breaking the night period with artificial light when the plant is about a month old. Continue the treatment until you want the plants to flower. (See the discussion of photoperiod in Chapter 3.)

Spring crops can be trimmed of buds when mature. The plant is left in the ground, and as the daylength increases, the plant will renew vegetative growth and flower once more in the fall. Plants can also be started in November or December indoors under lights and planted outdoors in February for harvest in April or May. The plants will grow faster under lights than they would outdoors under the weak winter sun. When they are placed outdoors, the long nights will induce flowering. By April the sunlight gets much stronger, perfect for flower development. Plants placed outdoors in February adjust easily to sunlight. Even so, they should be conditioned so that they do not suffer severe burn, as described in the Transplanting section in Chapter 14.

Rejuvenation

Plants grown in areas where the weather is mild can survive winters when there are no heavy freezes. During the winter the plants will grow very slowly, but as soon as the weather warms, and the light gets more intense, the plants respond. This technique can also be used to obtain a second growth crop during Indian summers. The second growth is not as vigorous as the original, but it does increase the total harvest.

To prepare plants for rejuvenation, leave three or four pairs of lower branches with leaves on the plant when you harvest. The leaves need not be large, but they must be green. Water and fertilize the plants. Within a few days the plants will show new growth.

The authors observed an outdoor container garden composed entirely of plants which survived a mild San Francisco Bay Area winter. These developed healthy second growth the following summer and flowered again in the fall. Some growers in Hawaii claim that their plants are three years old and that the plants have yielded as many as six crops of buds. Perennial marijuana plants also grow in Jamaica and Thailand.

Water Deprivation

Many cultivators begin to limit the amount of water their plants receive as soon as the flowers start to appear. Other growers give their plants as little water as possible after the middle of the plant's life. The plants are given small amounts of water only when they begin to wilt. (See Chapter 19 on the reasons for stressing the plants.)

Under water stress many of the leaves may die and fall from the plant. Sometimes the plants appear "burned," and turn brown or gold. At harvest, water-stressed plants may only have buds left on them and these may have the color, resin, and harshness typical of Colombian grass. These plants yield less grass at season's end. Not only are they smaller overall, but many of the leaves will have fallen away.

Water stress can be difficult to control in areas with heavy summer rain. Water-stressed plants often make up for their smaller size by a rapid burst of growth after a heavy rain. One method of control is to cover the ground with plastic sheets when it rains so that most of the water runs off.

Tacks and Nails

Some growers hammer nails or tacks into the stems of plants several weeks before harvest. Many growers use long thick nails; others prefer to use several half-inch-long tacks. The nails are usually placed at the base of the stem. This is supposed to "increase potency."

Figure 72. Wilted plant. Unless watered it will die.

Stem Splitting

This is a popular way to stress used by cultivators in the United States. The stem is split (not cut) at the base to form a space through the stem. Growers place a rock, small piece of wood, an old *Cannabis* stem, or piece of opium (in Africa) in the split. Sometimes the wound is bound with cloth or plastic. We don't recommend this procedure, and advise you to be careful not to kill the plants and ruin the harvest.

Varieties

Outdoor growers are well-advised to plant several varieties of marijuana, because some varieties adapt to their new environment better than others. Also, each variety (and to a small extent, each plant) has its own bouquet. By planting several varieties, cultivators assure themselves a varied selection of smoking material.

In areas with short growing seasons, many tropical varieties do not have a chance to flower. But immature material from these varieties may be more potent than mature flowers of a plant grown from seed of lower-quality grass. For instance, compare a flowering Mexican with a Colombian that doesn't. The Colombian may be better because the difference in varieties is so great. On the other hand, the Mexican may be better because it is flowering and has reached its full potential.

Intercropping

It is well-known that certain plants may be antagonistic to other species of plants, and that there are also beneficial relationships between species. *Cannabis* is known not to grow well among spinach.[222] Although tomatoes and tobacco have been recommended as crops to avoid when growing marijuana, because of pests and diseases that these plants may harbor,[67] marijuana grows very well in healthy tomato patches. Growers have also commented on how well marijuana grows when planted with corn, sugarcane, and beets.

Chapter Sixteen
Insects and Other Pests

Outdoors, where it functions as part of an ecological system, marijuana is less susceptible to insect attacks than it is indoors. In an outdoor environment, insects are subject to the vagaries of the weather, food supply, and predators. And marijuana grows so fast that insects usually do little damage. Plants, plant eaters, and predators usually maintain an equilibrium which minimizes damage. But this balance is disturbed by tilling and gardening, and may take a while to reestablish itself.

The soil surrounding your plants may be teeming with insects, and it would be unnatural not to see some on your plants. Most insects do not eat marijuana. The few that do are the food which helps to keep a small population of their predators alive. Insects in the garden need to be controlled only when there is a real threat of damage.

Marijuana is most vulnerable in its early stages. After the plant increases production of the cannabinoids and resins at the eighth or ninth week, most insects are repelled. When the plants are small, an occasional munch affects a relatively larger part of the plant. That same bite affects a relatively smaller part when the plant is larger.

The insects that infect marijuana indoors—aphids, mealy bugs, mites, and whiteflies—do best in humid conditions with constantly warm temperatures. Outdoors they rarely inflict much damage on marijuana. The pests that are most likely to damage marijuana are leafhoppers, treehoppers, cucumber beetles, thrips, flea beetles, several kinds of caterpillars, snails, and slugs. The younger the plants are, the more susceptible they are to attack. Your prime goal is to protect the plants during the first two vulnerable months. You need to keep the pest population low, so that the damage is relatively light. The pests don't have to be eliminated, only kept under control.

There are many ways to keep pests from damaging your crops. These fall into one or more of several categories: biological control; capture traps and barriers; home remedies; and chemical insecticides.

BIOLOGICAL CONTROL

The theory behind biological controls is that methods for control of pests can be found within nature. These methods are safer to humans and less damaging to the environment than commercial insecticides. Gardeners have many forms of biological control at their disposal, including companion planting, use of predators, and sprays made from plant extracts or ground-up insects.

Companion Planting

Some plants, including marijuana in its later stages, produce resins or essences which repel or kill plant pests. Some of them are general repellents that affect a broad range of plant pests; others affect specific species. Generally, the heavily scented plants, such as spices, mints, and other herbs, are most likely to have these qualities.

Some of the more familiar plants used to protect gardens are the Alliums, or onion family, with garlic, chives, green onions, and other oniony-type plants as members. This group repels a broad range of plant pests, such as aphids, spider mites, flea beetles, potato bugs, bean beetles, and many other insects, as well as rabbits and some deer. They are easily planted around the garden or between the marijuana plants. Just plant onion bulbs or the cloves from a garlic bulb so that the top of the bulb is about one inch deep. One garlic bulb yields quite a few cloves; so a large garden requires only a few bulbs.

Geraniums are reputed to repel leafhoppers and many kinds of beetles. These plants prefer a dry soil, thrive in full light, and usually grow two feet tall. Geraniums should be interspersed with the marijuana, or potted geraniums can be set out if problems develop. Tansy (*Tanaetum vulgare*) is a tall, fragrant, woody perennial which grows five feet tall. It protects against cutworms, beetles, cucumber beetles, and other eaters and borers.

Mints repel many insects and are sometimes used as mouse repellents. They are especially useful for the control of the flea beetle. They thrive in semi-shaded areas with rich soil.

Marigolds can be planted to eliminate nematodes. They are fast-growing annual plants which flower profusely. They come in many varieties, ranging in height from six to 30 inches. They grow in a wide range of soils and do best in the sun. The scented varieties—usually nonhybrids—offer the most protection.

All companion plants must be planted close to the plants to be protected, since their repellent qualities spread only a short distance beyond their circumference. They are effective when they are planted before the damage is apparent, and offer long-term protection. They are used when a pest is expected. For instance, growers in the San Francisco Bay Area expect rose leafhoppers to attack their plants. Since geraniums grow in the area as perennial plants, some growers plant them permanently in the garden. As the geraniums develop into small bushes, the hoppers leave, never to return.

Predators

Many of the insects in your garden are called beneficials, because they perform a useful service in the garden. Some of them eat decaying matter; others help in the pollination process; and some prey on insects which damage crops. Almost everyone is familiar with the ladybug, which eats aphids and insect eggs and has a voracious appetite. They are available commercially by the pint. The praying mantis eats slow-moving insects. When it first hatches, it starts out on aphids and mites. But as it grows larger, it eats bigger insects and worms. Mantis-egg cases are foam-like, straw-colored masses which contain 100 to 300 eggs. These cases are sold commercially but can also be found in the late fall in bushy areas. Another insect which is sold commercially as a plant protector is the green or brown lacewing. It has golden eyes, looks fragile, and flies erratically. But in their larval state, lacewings eat thrips, mites, caterpillar eggs, scale, leafhopper nymphs, aphids, and mealybugs. The trichogamma wasp is an egg parasite which lays its eggs in the eggs of over 200 species of insects, including many moths and butterflies which hatch into worm pests. *Cryptolaemus* is used to destroy mealybugs. Adults are released when mealybugs appear in the spring. They seek out the mealybug colonies and lay their eggs. When the eggs hatch the larvae wander around the infested area and eat the young mealybugs.

The use of commercially bred or gathered predators is most feasible in large gardens or fields. The insects may not have much effect on small gardens, since they wander off to find food and may never return. Try to buy from manufacturers who intentionally do not feed their product before shipping. Hungry predators are more likely to stay and eat the pests.

Insects are just one group of predators. Birds such as purple martins, robins, blue jays, chickadees, and even starlings and English sparrows eat large quantities of insects and other small pests. They can be attracted to the garden by placing a feeder, bird houses, and water in the area. When plants get larger, some gardeners let chickens, ducks, or geese run through the garden. In a short time, they pick it clean of pests and weeds. Reptiles and amphibians, including frogs, toads, snakes, lizards, and turtles, all eat garden pests and should be encouraged to make a home in the garden.

Homemade Repellents and Insecticides

Another way to control garden pests is to make sprays from plants which repel insects by using a juicer or blender or by making a tea. Ingredients can be found in most kitchens. Chile pepper, garlic, coffee, horseradish, radish, geranium, and tobacco are the usual mainstays of herbal sprays, although most strong-smelling herbs and spices have some repellent qualities. Many gardeners experiment to see what works in their garden. For instance, if an insect which bothers marijuana stays clear of a nearby weed, a tea or blended spray made from that plant may control the pest. But try it on only one plant (or part of a plant) first, because the spray may also be harmful to the marijuana.

Garlic is probably the most popular ingredient for general-purpose sprays made from kitchen ingredients. A typical formula is to soak three ounces of chopped or minced garlic in a covered container of mineral oil for a day. Then, slowly add a pint of lukewarm water in which a quarter ounce of real soap (Ivory will do) has been dissolved. Stir and let stand several hours, then strain. Use as a concentrate, adding between 20 to 100 parts water to one part concentrate.

Other recipes call for boiling the garlic or for grinding or juicing it. Some brewers add other spices to the basic formula. One recipe calls for one clove garlic, three cayenne peppers, one onion, a quarter ounce of soap, and sufficient water to blend.

Let it sit for three or four days before using, and use one part concentrate to 20 parts water. Homemade tobacco teas are sometimes used as insect sprays. Use one cigarette in a quart of water. Let it brew 24 hours before using.

Snails and slugs are attracted by yeast solutions, which are easily prepared from cooking yeast, sugar, and water. This is also why gardeners have success trapping these leaf munchers in bowls of stale beer. Place deep-sided containers at the soil level. The pests slide in and drown.

Gardeners should not overlook handpicking as a viable method of pest control. The foot or a quick thumb and forefinger can eliminate large numbers of pests and can keep a small garden pest-free. Collect the bugs and drop them in a tin can with some alcohol to kill them. Early morning is the best time to collect pests, since they are slower-moving until the sun warms them.

Snails, slugs, earwigs, and some other insects gather in cool, moist areas during the heat of the day. By providing just such a space in a garden, many of these pests can be located and destroyed. Place pieces of cardboard or boards around the garden; look under them each day.

Home Remedies

Gardeners and farmers have discovered and invented many ingenious ways to control insects without harming the environment. Some of the more popular ones are listed here, but there are many more, each suited to a particular situation.

Soap and water is an effective control measure for mealybugs, mites, leafhoppers (nymph stage), leaf miners, and aphids. Simply wash the plants thoroughly with a solution of two tablespoons of soap dissolved in a gallon of water. Rinse the soap off thoroughly. (Some growers feel that the addition of kerosene or alcohol makes the solution more effective, but these can harm the plants and dissolve THC.) This treatment does not eliminate all of the pests, and may need to be repeated weekly, but it does keep them under control.

Sprays are sometimes made from unhealthy insects, which are caught, ground up, and then sprayed back onto the plants. When the pests come in contact with the spray, they become infected with the pathogen and get sick. This method is very effective, and is considered safe, but it is not easy to capture sick insects. A variation on this technique was described in the October 1976 *Organic Gardening and Farming Magazine,* in which a spray was made from healthy insects. In a followup article in the May 1977 issue, the authors theorized that any population of insects contains pathogens. If enough insects are collected, some of them are sure to be sick, and they contain enough germs to spread the disease. To make an insect spray, capture about a hundred pests. (Make sure not to include any beneficial insects or the spray may also work against them.) Using a blender, mix them with a cup of spring water, strain, and dilute with enough water to spray your garden.

Whenever making or storing a spray, use a glass container. Metal or plastic ones may react with the chemicals that the liquids contain.

Another home remedy for the control of mites and aphids is a mixture consisting of a half cup of milk in four cups of wheat flour, added to five gallons of water. When it is sprayed on the undersides of the leaves, it suffocates the insects and then flakes off as it dries.

Some growers use mulches to control insects. Cedar chips repel beetles, moths, mites, and mealybugs. Aluminum foil is used for aphid and thrip control on small plants; the reflected light disorients them and they do not land on the plants. A sprinkling of cream of tartar eliminates ants, and boric acid kills roaches. Sulfur powders, available at nurseries, are used to control mites and fungus infections.

Organic Insecticides

Pyrethrum, rotenone, and ryania are effective insecticides which come as powders (dusts) or sprays. They are concentrated forms of naturally occurring plant substances, and are considered harmless to warm-blooded animals when used as directed.

Ryania, which is found in the roots of a tropical shrub, is most effective against chewing insects, worms, and larvae, which it incapacitates, rather than kills.

Rotenone is a general-purpose insecticide with little residual effect; that is, it breaks down soon after application, and is therefore one of the safest insecticides. Two or three dustings during the seedling stages afford protection against most insects and bugs.

Pyrethrum is one of the most powerful natural insecticides,

and is effective against a wide range of pests. It is also relatively nontoxic to bees and ladybugs. Pyrethrums are found in the pyrethrum plant as well as in chrysanthemums. They are non-persistent, and in small doses may make the insects sick without killing them. These insecticides are available at many nurseries and may provide the surest, easiest form of protection against serious insect attack.

Barriers and Traps

In gardens and small farms, insects and other pests are some-times controlled by the use of traps and barriers that prevent them from reaching the marijuana. When the plants are young, they can be protected from cutworms, caterpillars, snails, and slugs by a collar that is buried an inch into the ground and is six inches high. Some growers face it with aluminum foil, which many insects seem to dislike. One ingenious grower painted collars with molasses to capture the crawlers. She also caught a significant number of leafhoppers. Commercial stickums such as Tanglefoot can also be used to trap insects.

Snails, slugs, and some crawling insects are repelled by a border perimeter of lime, potash (wood ash), sulfur, sharp sand, or cinders. Place a thin layer, six inches wide, around the peri-meter of the garden, or around each plant. Flea beetles and some other flying insects are repelled by wood ashes dusted on the leaves. The powders are water-soluble; so they should be replaced after a heavy rain. Crawling pests sometimes have a hard time reaching plants grown in containers or raised beds.

Flying insects, such as leaf and treehoppers, can be prevented from getting to plants by barriers made from cheesecloth. Other growers place cardboard sticky with glue between plants, and then shake the plants. The cardboard catches a good proportion of them. One innovative grower in Palo Alto, California, placed a furniture crate, with the top cut off and with Tanglefoot spread on the inside, around each of his six plants. He said that by shaking the plants, he eliminated leafhoppers in four days.

CHEMICAL INSECTICIDES

Insecticides were developed as an easy way to control pests. They have an immediate dramatic effect, but the long-range damage that they do to the entire ecological system is sometimes overlooked. The chlorinated hydrocarbons, such as DDT, DDC, Aldrin, Kelthane, and Dieldrin, were the most dangerous commercial insecticides. They affect warm-blooded animals and are no longer available. (In no case should any of these be used.)

Diazinon, Sevin, and Malathion are three insecticides which are often sold in nurseries to protect vegetable crops. They are considered safe for warm-blooded animals and have a limited residual effect, since they break down in a few days. But these insecticides are not too selective and may kill beneficials as well as pests. Sevin is the most toxic and kills the widest range of insects, including bees.

These chemicals come as sprays, powders, and baits, formulated for specific pests. They should be used only when an intolerable situation has developed. Plants should be harvested only after the required safety period has passed since application. This period is from two to 35 days, and is specifically listed on all insecticides that can be safely used. Insecticides should be used and handled carefully, following instructions, wearing protective clothing, with no children or pets around. It is advisable to use a mask when applying dusts and to work upwind.

COMMON PESTS

Cucumber Beetles

Cucumber beetles are about a quarter-inch long and look a lot like ladybugs. There are several species of cucumber beetles. The striped beetle is found east of the Rocky Mountains. It is yellow, has two or three black stripes running down its back, and has a black head. The spotted cucumber beetle has a yellow-green back with 11 or 12 black spots and a black head. There are related species, such as the banded cucumber beetle, throughout the United States. The larvae of all varieties are white, turning brownish at the ends, slender, about one-third inch long.

Cucumber beetles do the most damage in the early spring, when the adults come out of hibernation and begin to eat the new growth and leaves. These leaf-eating adults damage young marijuana, especially when there is a scarcity of other food. They also transmit bacterial diseases and viruses to the plants.

Within a few weeks after they come out of hibernation, they lay their eggs at the base of plant roots. The larvae bore into the roots, where they mature by mid-summer. The larvae of the striped cucumber beetle feed only on melon- and cucumber-type plant roots. The spotted-beetle larvae are fond of corn, and are known as the "Southern cornroot worm" in some places.

The best way to prevent cucumber-beetle attacks is to keep the areas that you plant isolated from corn and melon plantings. Heavy mulching or tilling destroys the pests when they are hibernating. Late plantings minimize damage inflicted by cucumber beetles.

Cucumber beetles can be controlled by use of Rotenone or Malathion. Dust several times during seedling growth. These beetles are also prey to many insects, including the common garden soldier beetle, predator flies, wasps, and nematodes. Hand picking is also an effective control for cucumber beetles.

Thrips

Thrips are slender, yellow or brownish, winged insects about 1/25 inch long. They have fragile wings which keep them aloft while they are blown by the wind. Thrips have a cone-shaped mouthpart, which they use to cut stems in order to suck plant juices. The larvae look like adults, but are smaller and wingless. Most thrips feed on a range of plants, especially onion and other bulbs, and marijuana is at most a marginal part of their diet. A well-cultivated marijuana plant can outgrow any damage that thrips are likely to inflict.

Thrips hibernate in plant debris during the winter and begin sucking in early spring. They lay eggs during warm weather, and can produce a new generation every two weeks. Since thrips eat a varied diet, keeping the garden area clear of weeds is an effective control. Thrips can also be controlled by turning debris under, so that their nesting sites are destroyed.

Thrips can be controlled by use of tobacco sprays, Rotenone, or Malathion. Aluminum-foil mulches are effective thrip repellents. The light reflected from the foil confuses their sense of direction.

Flea Beetles

There are many species of flea beetles. The adults range in size between one-twentieth and one-fifth of an inch, and are usually black or metallic green or blue. They are called flea beetles because they use their enlarged hind legs to jump like fleas when disturbed. Many flea beetles are host-specific, and probably only a few species munch on marijuana.

Flea beetles hibernate in plant debris. By plowing the debris under, their hibernation places are eliminated, and there should be few pests the following spring. Flea beetles are repelled by a mixture of equal parts of wood ashes and limestone sprinkled on foliage every few days. Containers of the mixture may also be placed around the plants. Garlic sprays also repel flea beetles. The chemical poisons used specifically for flea beetles are stomach poisons, which break down slowly and may not be safe to inhale. Home remedies are best for flea beetles.

VERTEBRATE PESTS

Mammals

Until it develops a hard fibrous main stem, usually at about two months, the young marijuana plant attracts rodents, including mice, rabbits, moles, squirrels, groundhogs, and rats, as well as raccoons. Cats are probably the best means of rodent control. They stalk small prey, go after any movement, and are active at night, when most of these animals forage. Young plants are often protected from rodents by placing a coffee can with top and bottom removed around each plant. When the plants get bigger, they can be protected from rabbits and other animals with a wire fence three feet in height. A double layer of one-inch chicken wire is most effective. But many animals can climb or burrow; so more ingenious methods are needed to protect the plants. Rodents, especially moles, are repulsed by castor beans and castor oil. A formula that gardeners sometimes use is two parts castor oil, one part detergent, mixed to a consistency of shaving cream in a blender. Use a tablespoon of concentrate per gallon of water. Spray or mist the solution on the plants.

Rabbits shy away from blood, bloodmeal, and tankage. To use, sprinkle the powder around the perimeter of the plot in a band about a foot wide. They can also be mixed into a concentrated solution and applied as a spray. However, the smell of blood may attract mongoose or other predators, which dig up

Figure 73. Tin cans protect against cutworms and many other plant eaters.

the garden in search of flesh. Noise from radios, chimes, and bells deter some animals, and human smells such as hair and urine may also deter some animals. In dry areas, a half-filled bucket of water is an effective rodent trap. The animals fall in and drown.

Deer seem to go out of their way to munch on tender marijuana leaves, but generally don't bother marijuana after it has grown for a few months. Gardeners and farmers use many ingenious techniques to keep them away from crops. Sturdy fences are the best deterrent. The fences should be about 10 feet high: the bottom five feet should be made up of single strands of wire string at two-foot intervals. The wire strands prevent deer from jumping the fence. Some growers use fresh blood, dried blood, or bloodmeal to deter them, placing it in either powder or liquid form around the perimeter of the garden. Other growers claim that human hair, or manure from predators such as wolves, bears, lions, and even dogs, keeps them out. Lion urine (glans extract) is available commercially, and is said to be an effective deterrent against many animals.

Figure 74. Fat rat munching marijuana. *Photo by A. Karger.*

Figure 75. Problem solved; or, never underestimate the power of a peanut. *Photo by A. Karger.*

Birds

On the whole, birds are beneficial, rather than harmful, in the garden. Most of the common species, including English sparrows, robins, swallows, wrens, finches, bluejays, bluebirds, and starlings, eat insects and other garden pests as a substantial portion of their diet. The only time that birds may be harmful is during planting, when they sometimes feed on the planted seed. The main culprits seem to be starlings, sparrows, and crows. They can be kept off the planted areas physically, by means of plastic netting or fencing, which is sold commercially for the purpose, or by using scarecrows, aluminum strips, or noise makers. Once the plants have germinated, birds are no longer a threat and should be encouraged to nest in the area, since they are an ideal biological control for plant pests.

PART 4.
Flowering, Breeding, and Propagation

Chapter Seventeen
Genetics and Sex in *Cannabis*

Sex is an inherited trait in *Cannabis,* and can be explained in much the same terms as human sexuality can. Like a human being, *Cannabis* is a diploid organism: its chromosomes come in pairs. Chromosomes are microscopic structures within the cells on which the genes are aligned. *Cannabis* has 10 pairs of chromosomes ($n = 10$), for a total of 20 chromosomes ($2n = 20$).

One pair of chromosomes carries the primary genes that determine sex. These chromosomes are labeled either X or Y. Male plants have an XY pair of sex chromosomes. Females have XX. Each parent contributes one set of 10 chromosomes, which includes one sex chromosome, to the embryo. The sex chromosome carried by the female ovule can only be X. The one carried by pollen of the male plant may be either X or Y. From the pollen, the embryo has a 50/50 chance of receiving an X, likewise for Y; hence, male and female progeny appear in equal numbers (in humans, the sperm carries either an X or a Y chromosome).

FLOWERING

Male Plant

Under natural light, males usually start to flower from one to four weeks before the females. Where the photoperiod is artificially controlled, as with electric lights, males respond quickly (in about a week) to a change to short photoperiods and usually show flowers sooner than the females.

Male flowers develop quickly, in about one to two weeks on a vigorous plant, but not uniformly. Scattered flowers may open a week or more before and after the general flowering, extending the flowering stage to about four weeks.

The flowering stage continues to demonstrate the male's tall, relatively sparse growth. Most of the flowers develop near the top of the plant, well above the shorter females. The immature flower buds first appear at the tips of the main stem and branches. Then tiny branches sprout from the leaf axils, bearing smaller clusters of flowers. The immature male flowers are closed, usually green, and develop in tight clusters of knob-like buds. The main parts of the male flower are five petal-like sepals which enclose the sexual organs. As each flower matures, the sepals open in a radiating pattern to reveal five pendulous anthers (stamens).

Inside the ovoid, sac-shaped anthers, pollen grains develop. Initially, pollen sifts through two pores near the top of the anther; then, starting from the pores, longitudinal slits slowly open (zipperlike) over the course of a day, releasing pollen to the wind. Once a flower sheds pollen, it shortly dies and falls from the plant. Pollen release is the final act in the life cycle of the male plant. Normally, male plants begin to die one to two weeks after the bulk of their flowers have shed pollen. Healthy males may continue to flower for several more weeks, but secondary growth seldom has the vigor of the initial bloom.

Female Plant

The female plant generally starts to flower later than the male, under either natural light or an artificially controlled photoperiod. Female marijuana plants flower when the average daily photoperiod is less than about 12 to 13 hours. However, some varieties and individuals may flower with a photoperiod of over 14 hours. Some Colombian varieties may not respond until the photoperiod falls below 12 hours for a period of up to three weeks.

The duration of flowering also depends on the particular rhythm of the variety, as well as growing conditions, and whether or not the plant is pollinated. Within these variables, females maintain vigorous growth and continue to rapidly form flowers for a period that ranges from 10 days to about eight weeks.

Females generally do not grow much taller during flowering. Growth emphasizes a "filling out," as flower clusters develop from each leaf axil and growing tip. Normally, the flowers arise in pairs, but the pairs form tight clusters of 10 to over 100 individual flowers that are interspersed with small leaves.

Figure 76. Female in full bloom. *Photo by G. Bratt.*

These clusters are the "buds" of commercial marijuana. Along the top of the main stem and vigorous branches, "buds" may form so thickly that the last foot or more of stem is completely covered. Usually the leaves that accompany the flowers tend toward simpler structure, until each leaf has one to three blades.

The visible parts of the female flower are two upraised stigmas, one-quarter to one-half inch long, usually white or cream, sometimes tinged with red, that protrude from a tiny, green, pod-shaped structure called the floral bract. This consists of modified leaves (bracts and bractioles) which envelop the ovule or potential seed. The mature bract is a tiny structure, about 1/8 inch across and 1/4 inch long. When fertilized, a single seed begins to develop within the bract, which then swells until it is split by the mature seed.

Bracts are covered more densely with large resin glands than is any other part of the plant, and are the most potent part of the harvest. Resin glands may also be seen on the small leaves that are interspersed among the flowers.

The differences between male and female *Cannabis* become more apparent as the plants mature. The same can be said of the differences between varieties. Often, two varieties may appear to be similar, until they actually flower and fill out to different forms. These appear in many ways: some varieties maintain opposite phyllotaxy with long internodes throughout flowering; bud sizes vary from about one-half inch to about three inches,

Figure 77. *Upper left:* Buds form thickly into colas along the top of the main stem and branches (full bloom). *Upper right:* A cola about two feet long. *Lower left:* A huge leafy cola. *Lower right:* Long, slim buds form late in the year when light is weak. (These four colas are from Mexican plants.)

with a norm of about one to two inches; buds may be tightly arranged along the stem, yielding a "cola" two feet long and four inches thick; and some varieties only form buds along their main stem and branch tips, with few "buds" forming along the branches.

When a female is well-pollinated, growth slows and the plant's energy goes into forming seeds and thus into the continuation of the species. Some plants (but only the more vigorous ones) will renew flowering even when pollinated. Females that are not well-pollinated continue to form flowers rapidly. This extends the normal flowering period, of 10 days to four weeks, up to eight weeks or more.

Individual flowers are pollinated by individual pollen grains. In a matter of minutes from its landing on a stigma, the pollen grain begins to grow a microscopic tube, which penetrates the stigma and reaches the awaiting ovule wrapped within the bracts. The pollen tube is a passageway for the male's genetic contribution to the formation of the embryo (seed).

The union of the male and female complements of genes completes fertilization and initiates seed formation. The stigmas, having served their purpose, shrivel and die, turning rust or brown colors. On a vigorous female, the seeds reach maturity in about 10 days. When growing conditions are poor, the seed may take five weeks to ripen to full size and color. Naturally, all the flowers do not form, nor are they pollinated at the same time—and there will be seeds that reach maturity weeks before others do. Although each flower must be individually fertilized to produce a seed, a single male plant can release many millions of pollen grains. A large female plant can produce over 10,000 seeds.

SEXUAL VARIANTS IN *CANNABIS*

Cannabis has been studied for many years because of its unusual sexuality. Besides the normal dioecious pattern, where each plant bears exclusively male or female flowers, it is not uncommon for some plants to have both male and female flowers. These are called hermaphrodites, or monoecious plants, or intersexes. Hermaphroditic plants form normal flowers of both sexes in a wide variety of arrangements, in both random and uniform distributions.

Natural Hermaphrodites

Some hermaphrodites seem to be genetically determined (protogenous). That is, they naturally form flowers of both sexes given normal growing conditions. Possibly genes carried on the autosomes (the chromosomes other than the sex chromosomes) modify the normal sexual expression. Monoecious varieties have been developed by hemp breeders in order to ensure uniform harvests.

It is also possible that these particular plants are polyploid, which means they have more than the usual two sets of chromosomes. This kind of hermaphrodite may have XXY (triploid), or XXYY or XXXY (tetraploid) sex chromosomes. However, no naturally occurring polyploids have ever been verified (by observation of the chromosomes) in any population of *Cannabis*. Polyploids have been induced in *Cannabis* by using mutagens, such as the alkaloid colchicine.

Whatever the genetic explanation may be, one or more of these natural hermaphrodites may randomly appear in any garden. They are sometimes faster-maturing, have larger leaves, and are larger in overall size than their unisexual siblings. They usually form flowers of both sexes uniformly in time and distribution, and in some unusual patterns. For example, from Mexican seed, we have seen a plant on which each separate flowering cluster consisted of both female and male flowers: an upper section of female flowers had upraised stigmas, and a lower section of male flowers dangled beneath the female flowers. In other plants from Mexican seed, the growing tips throughout the plant have female flowers; male flowers sprout from the leaf axils along the main stem and branches. Plants from "Thai" seed sometimes form male and female flowers on separate branches. Branches with female flowers tend to predominate, but branches having mostly male flowers are located throughout the plant.

Abnormal Flowers, Intersexes, Reversals

Gender is set in the new plant at the time of fertilization by its inheritance of either the X or the Y chromosome from the male

Figure 78. *Upper left:* Abnormal flowers. *Lower left:* Male flowers on a female plant. *Upper right:* Sexes on separate branches. *Lower right:* Male flower in female bud (reversing).

(staminate) plant. With germination of the seed, the environment comes into play. Heritage sets the genetic program, but the environment can influence how the program runs. (Sexual expression in *Cannabis* is delicately balanced between the two.) The photoperiod, for example, controls the plant's sequence of development. Also, the plant's metabolism and life processes are dependent on growing conditions. When the environment does not allow a balance to be maintained, the normal genetic program may not be followed. This is mirrored by abnormal growth or sexual expression.

Abnormal Flowers   Abnormal sexual expression includes a whole range of possibilities. Individual flowers may form abnormally, and may contain varying degrees of both male and female flower parts. For instance, a male flower may bear a stigma; or an anther may protrude from the bracts of a female flower. Abnormally formed flowers are not often seen on healthy plants, although if one looks hard enough, a few may be found in most crops. When many of the flowers are abnormal, an improper photoperiod (coupled with poor health) is the most likely cause. Abnormal flowers sometimes form on marijuana grown out of season, such as with winter or spring crops grown under natural light.

Intersexes and Reversals   Much more common than abnormally formed flowers is for the plant's sex to be confused. One may find an isolated male flower or two; or there may be many clusters of male flowers on an otherwise female plant, or vice versa. These plants are called intersexes (also hermaphrodites or monoecious plants). Intersexes due to environmental causes differ from natural hermaphrodites in having random distributions and proportions of male and female flowers. In more extreme cases, a plant may completely reverse sex. For example, a female may flower normally for several weeks, then put forth new, sparse growth, typical of the male, on which male flowers develop. The complete reversal from male flowering to female flowering also happens.

All other things being equal, the potency of intersexes and reversed plants is usually less than that of normal plants. If there are reversals or intersexes, both of the sexes will usually be affected. Female plants that reverse to male flowering show the biggest decline. Not only is the grass less potent, but the amount of marijuana harvested from male flowers is neglible compared to the amount of marijuana that can be harvested from a normal female. Plants that change from male to female flowering usually increase their potency, because of the growth of female flower bracts with their higher concentration of resin. Female flowers on male plants seldom form as thickly or vigorously as on a normal female. Between the loss in potency and the loss in yield because of females changing to males, a crop from such plants is usually inferior, in both yield and potency, to one from normal plants.

Environmental Effects

Many environmental factors can cause intersexes and sexual reversals. These include photoperiod, low light intensity, applications of ultraviolet light, low temperatures, mutilation or severe pruning, nutrient imbalances or deficiencies, senescence (old age), and applications of various chemicals (see bibliography on sex determination).

The photoperiod (or time of planting using natural light) is the most important factor to consider for normal flowering. In 1931, J. Schaffner[105] showed that the percentage of hemp plants that had confused sexual characteristics depended on the time of year they were planted. Normal flowering (less than five percent of the plants are intersexes) occurred when the seeds were sown in May, June, or July, the months when the photoperiod is longest and light intensity is strongest. When planted sooner or later in the year, the percentage of intersexuals increased steadily, until about 90 percent of the plants were intersexual when planted during November or early December.

Marijuana plants need more time to develop than hemp plants at latitudes in the United States. Considering potency, size, and normal flowering, the best time to sow for the summer crop is during the month of April. Farmers in the south could start the plants as late as June and still expect fully developed plants.

If artificial light is used, the length of the photoperiod can influence sexual expression. Normal flowering, with about equal numbers of male and female plants, seems to occur when the photoperiod is from 15 to 17 hours of light for a period of three to five months. The photoperiod is then shortened to 12 hours to induce flowering. With longer photoperiods, from 18 to 24 hours a day, the ratio of males to females changes, depending on whether flowering is induced earlier or later in the plant's life. When the plants are grown with long photoperiods for six months or more, usually there are at least 10 percent more male than female plants. When flowering is induced within three months of age, more females develop. Actually, the "extra" males or females are reversed plants, but the reversals occur before the plants flower in their natural genders.

Some plants will flower normally without a cutting of the photoperiod. But more often, females will not form thick buds unless the light cycle is cut to a period of 12 hours duration. Don't make the light cycle any shorter than 12 hours, unless the females have not shown flowers after three weeks of 12-hour days. Then cut the light cycle to 11 hours. Flowers should appear in about one week.

Anytime the light cycle is cut to less than 11 hours. some intersexes or reversed plants usually develop. This fact leads to a procedure for increasing the numbers of female flowers indoors. The crops can be grown for three months under a long photoperiod (18 or more hours of light). The light cycle is then cut to 10 hours. Although the harvest is young (about five months) there will be many more female flower buds than with normal flowering. More plants will develop female flowers initially, and male plants usually reverse to females after a few weeks of flowering.

Of the other environmental factors that can affect sexual expression in *Cannabis,* none are as predictable as the photoperiod. Factors such as nutrients or pruning affect the plant's overall health and metabolism, and can be dealt with by two general thoughts. First, good growing conditions lead to healthy plants and normal flowering: female and male plants occur in about equal numbers, with few (if any) intersexes or reversed plants. Poor growing conditions lead to reduced health and vigor, and oftentimes to confused sex in the adult plant. Second, the age of the plants seems to influence reversals. Male plants often show female flowers when the plant is young (vigorous) during flowering. Females seven or more months old (weaker) often develop male flowers after flowering normally for a few weeks.

Anytime the plant's normal growth pattern is disrupted,

normal flowering may be affected. For instance, plants propagated from cuttings sometimes reverse sex, as do those grown for more than one season.

SEXING THE PLANTS

The female plant is more desirable than the male for marijuana cultivation. The female flowering clusters (buds) are usually the most potent parts of the harvest. Also, given room to develop, a female generally will yield twice as much marijuana as her male counterpart. More of her weight consists of top-quality buds.

Because the female yields marijuana in greater quantity and quality, the earlier you can discover each plant's gender, the sooner you can devote your attention to nurturing the females. Where space is limited, such as in indoor gardens and small outdoor plots, most growers prefer to remove the males as soon as possible, and leave all available space for the females. To harvest sinsemilla (seedless female buds), you must remove the male plants before they mature and release pollen.

Differences in the appearance of male and female *Cannabis* become more apparent toward maturation. During the seedling stage, gender is virtually impossible to distinguish, although in some varieties the male seedling may appear slightly taller and may develop more quickly.

We know of no way to discover gender with any certainty until each plant actually forms either pollen-bearing male flowers or seed-bearing female flowers. However, certain general characteristics may help. Using guidelines like the following, growers who are familiar with a particular variety can often predict gender fairly accurately by the middle stage of the plant's life.

Early Vegetative Growth

After the initial seedling stage, female plants generally develop more complex branching than the male. The male is usually slightly taller and less branched. (Under artificial light, the differences in height and branching are less apparent throughout growth.)

Some plants develop a marked swelling at the nodes, which is more common and pronounced on female plants.

Middle Vegetative Growth

In the second to fourth months of growth, plants commonly form a few isolated flowers long before the actual flowering stage begins. These premature flowers are most often found between the eighth and twelfth nodes on the main stem. Often they appear near each stipule (leaf spur) on several successive nodes, at a distance two to six nodes below the growing tip. These individual flowers may not develop fully and are often hard to distinguish as male or female flowers. The fuzzy white stigmas of the female flower may not appear, and the male flower seldom opens but remains a tightly closed knob. However, the male flower differs from the female; it is raised on a tiny stalk, and the knob is symmetrical. The female flower appears stalkless and more leaflike.

The presence of premature female flowers does not assure that the plant is a female, but premature male flowers almost always indicate a male plant. Unfortunately, it is much less common for male plants to develop premature male flowers than for female flowers to appear on either plant. For example, in one garden of 25 mixed-variety plants, by age 14 weeks, 15 plants showed well-formed, premature female flowers with raised stigmas. Eight of these plants matured into females and seven became males. Only two plants showed premature male flowers and both of these developed into males. The eight remaining plants did not develop premature flowers or otherwise distinguishable organs until the actual flowering stage at the age of 21 weeks. From these eight, there were four females, three males, and one plant bearing both male and female flowers (hermaphrodite). It does seem, however, that plants bearing well-formed female flowers, on several successive nodes, usually turn out to be females.

Preflowering

In the week or two prior to flowering and throughout flowering, many common marijuana varieties follow two general growth patterns which depend on gender. With these varieties, you can tell gender by the spacing between the leaves (internodes). For

Figure 79. Premature flowers are found on the main stem next to the leaf spurs. *Upper left:* Early female flower without stigmas. *Lower left:* Undifferentiated (indistinguishable). *Center:* Early male flower. *Upper and lower right:* Well-formed female flowers on successive nodes usually indicate a female.

the female, the emphasis is on compact growth. Each new leaf grows closer to the last, until the top of the plant is obscured by tightly knit leaves. The male elongates just prior to showing flowers. New growth is spaced well apart and raises the male to a taller stature. This may be the first time the male shows its classic tall, loosely arranged profile.

SINSEMILLA

Sinsemilla* is any marijuana consisting of seedless female flower buds. Sinsemilla is not a variety of marijuana; it is the seedless condition that results when the female flowers are not fertilized with pollen.

In the United States, most sinsemilla comes in the form of Thai sticks that are imported from Southeast Asia and Japan. Thai sticks are made up of seedless buds wrapped around a sliver of bamboo or a long wooden matchstick. The buds, which may be on one or more stems, are secured with a hemp fiber wound around the stick. A growing amount of fine sinsemilla now comes from domestic sources, such as Hawaii and California. The grass is usually boxed or bagged with pure buds that are manicured (extraneous leaf removed). Infrequently sinsemilla comes from Mexico and, rarely, from Colombia.

Sinsemilla has a reputation as high-potency marijuana, with a sweet taste and mild smoke. It doesn't have the harsh, gagging qualities of the usual Colombian and Mexican grasses. These qualities, however, have nothing to do with sinsemilla as such. The potency of any grass depends primarily on the variety and development of the plant, and the taste and mildness of the smoke depend on the condition of the plant when harvested and the cure. Heavily seeded grass can be as mild and sweet-smoking as sinsemilla when it is properly handled.

When buying grass, remember that sinsemilla indicates a conscientious effort on the grower's part to bring you the best possible product. Sinsemilla is almost pure smoking material

---

*The word "sinsemilla" comes from the Spanish, and means "without seeds." It is also spelled "sansimilla."

Figure 80. Thai sticks.

with no wasted weight in seeds. An ounce of sinsemilla has about twice as much smoking material as a typical seeded ounce. Also, any marijuana that is fresh, with intact buds, indicates less deterioration of cannabinoids.

Sinsemilla is becoming a preferred form of grass with home-growers, many of whom believe that a seedless female is more potent than a seeded one, reasoning that the plant's energy goes to the production of resin rather than seed. There seem to be no scientific studies on this point. Many experienced growers believe the difference is small, perhaps 10 percent.

From observing the resin glands on the bracts, one sees that they continue to develop in size after pollination. Any difference from the unseeded state is not apparent. Whether pollination does in fact hamper or lessen resin production or potency is questionable. But the effect on the plant as a whole can be dramatic. Usually when the female is well-pollinated, growth noticeably slows, and the plant enters the last phase of life, which is seed set. Seed set is a period of incubation, in which the seeds grow and reach their mature state. New growth forms more slowly and lacks the vitality of the bloom before pollination. The plant's reaction to pollination is relative. The more thoroughly pollinated the female is, the more pronounced the change in rhythm from vigorous growth to incubation. A plant on which only a few flowers have been fertilized continues to actively form flowers as sinsemilla.

Not all plants react alike to pollination. When the weather is good and the plant vigorous, even a well-seeded plant may bloom a second or third time before the rate of growth starts a final decline.

To put this in perspective, the main advantage to growing sinsemilla is that the plant remains in a flowering state for a longer period of time. Flowers may rapidly form for four to ten weeks. The flower buds develop larger and more thickly along the stems, yielding more top-quality grass (more buds) than in the seeded condition.

Anyone can grow sinsemilla. Simply remove the male plants before they release pollen. Given a normal spring planting, males usually flower in August and September, but may begin to flower as early as mid-July. Under artificial lights, males sometimes flower after only three months, and before the grower has shortened the photoperiod. Even though the females are not flowering, remove the males from the room before any flowers open. Indoors, the pollen will collect as dust and can fertilize the females weeks later.

Male flowers mature quickly, in about one to two weeks after the immature buds are first visible. Check each plant about twice a week to make sure you harvest all the males before any shed pollen. If you can't visit your garden consistently, then thin the garden, using the preceding section on "Sexing" as a guide. Even though you may not get all the males, the females will be more lightly seeded. Actually, even in carefully watched gardens, the females may have a few seeds. Pollination may come from an occasional male flower on a basically female plant, or a female may reverse and form male flowers. And pollen may come from a neighbor's garden, a problem that is becoming more common. But in practical terms, an occasional seed makes no difference. The female can form thousands of flowers, and when only a few are pollinated, there is little impact on the plant's growth.

Chapter Eighteen
Propagation and Breeding

PRODUCING SEEDS

Marijuana is naturally prolific. It has been estimated that a single male plant can produce over 500 million pollen grains.[41] A large female plant can bear tens of thousands of seeds. In nature, pollen is carried from the male flowers to the stigmas of the female flowers by air currents or the wind. Indoors or out, if the plants are simply left on their own, most gardens produce many more seeds than are needed for the next crop.

Seeds usually become viable within two weeks after pollination, although they may not have developed good color by this time. The color can take several more weeks to develop, particularly indoors or late in the year, when the light is not as strong. Once seeds are plump, well-formed, and of a mature size, most of them will be viable. When seeds have also developed good color, their viability should be over 90 percent.

Pollination may also be carried out artificially. Pollen can be collected and then transferred to the female flowers with a cotton swab or artist's brush, or shaken directly over the flowers. Store pollen in a clean, open container and keep in a dry area at moderate temperature. Remove any flowers or vegetative matter from the pollen, because they encourage fungal attack.

One advantage of artificial pollination is that only the flowers on certain plants need be pollinated. This allows you to harvest most of your grass as sinsemilla, while developing seed on part of the plant. If you have only a few plants, pollinate a single branch, or perhaps only a few lower buds, in order to leave the most potent buds seedless.

A good way to insure a thorough pollination, and to avoid contaminating other females, is to loosely tie a transparent bag containing pollen directly over individual buds, branches, or whole plants. Shake the bag to distribute the pollen and carefully remove it from several hours to a few days later.

To avoid contaminating a sinsemilla crop, you must remove any males from the garden before their flowers open. Males in pots can simply be moved to another area or room if you want to keep them growing. Male plants can complete development even in low light; so they do not need artificial light. Otherwise, the best procedure is to harvest the males intact by cutting them at their base after some flowers have formed distinct (but unopened) buds. Hang the whole plants upside down in a sheltered area where there is moderate light and where temperatures and humidity are not extreme. Place clean plates or sheet plastic beneath the plants to catch falling pollen. Generally there is enough stored water in the plant for the unopened flowers to mature and drop pollen. Well-formed flowers may open the next day. Usually all the flowers that are going to open will do so within two weeks.

Pollen gradually loses viability with time, but pollen that is about three weeks old generally has sufficient viability for good seed production. However, the age of the pollen may influence the sex ratio of the next generation.

For instance, in a 1961 study with hemp plants,[97] the percentage of females in the next generation was 20 percent higher than in the control plants (natural pollination) when pollen 14 to 17 days old was used. A small increase in female-to-male ratios also occurred when pollen was fresh (six hours or less). The age of the stigmas appeared not to affect the sex ratio.

PRODUCING FEMALE SEEDS

If it were possible to know which seeds are female and which are male, marijuana growing would be even simpler than it is. There is no practical way to discern the gender of a seed—but there is a simple procedure for producing seeds that will all grow into female plants.

To produce female seeds, the plants are fertilized with pollen from male flowers that appear on a basically female plant. Such flowers appear on intersexes, reversed females, and hermaphrodites (see Chapter 17). Female plants have an XX complement of sex chromosomes; therefore, the pollen from male flowers that form on female plants can only carry an X chromosome. All seeds produced from flowers fertilized with this "fe-

Figure 81. Harvest males when flowers are well-
formed but not yet open.

male" pollen will thus have an XX pair of sex chromosomes,
which is the female genotype.

Although the male *Cannabis* plant can produce female flowers, it cannot produce seed; so there is no chance of mistakenly
producing seed on a male plant. It is possible to use pollen from
an intersexual plant that is basically male (XY); the resulting
crop of seeds will have the normal 1:1 ratio of males to females.
For this reason, choose a plant that is distinctly female as a
pollen source. A female plant with a few random male-flower
clusters, or a female plant that has reversed sex are both good
pollen sources. The seed bearer can be any female, female inter-
sex, or reversed-female plant.

In most crops, careful inspection of all the females usually
reveals a few male flowers. And often, when females are left
flowering for an extended period of time, some male flowers
will develop. If no male flowers form, you can help to induce
male flowers on female plants by severe pruning. One such procedure is to take the bulk of the harvest, but to leave behind
some green leaves to maintain growth (as described in the section on "Double Harvests" in Chapter 20). Most of the plants
will continue to form female flowers, but male flowers are also
likely to form. At times, the plants may not grow particularly
well, and may in fact form distorted and twisted leaves, but
they will produce viable seeds as long as some stigmas were
white when pollinated. (Remember, it only takes a few fertile

Figure 82. A solitary male flower on a female plant provides "female" pollen. (Also see Figure 84 for a female plant reversing sex.)

Figure 83. Growth may not be vigorous, but seeds will form if stigmas are white when pollinated.

buds to produce hundreds of seeds.) Pollinate the female flowers by hand as soon as pollen becomes available.

Under artificial lights, turn the light cycle down to eight hours after cutting the plants back. The short cycle helps to induce male flowers on female plants.

Male-free seed can also be produced by pollen from natural hermaphrodites. The progeny, however, may inherit the hermaphroditic trait, resulting in a crop with some hermaphrodites as well as females. This could be a problem if you want to grow sinsemilla the next crop.

BREEDING

Breeding *Cannabis* is done simply by selecting certain plants to be the pollinators and the seed bearers. Characteristics such as fast growth, early maturation, and high potency might be the reasons for choosing one plant over another. Selection can be by means of the male plants, the females, or both. A simple procedure would be to harvest all male plants, sample each for potency, and use the most potent plant for the pollen source. At harvest, compare the seeded females for potency, and use seeds from the most potent plant for the following generation.

There are two basic approaches to breeding. One is inbreeding, and the other is outbreeding. Inbreeding involves starting with a single variety and crossing individuals to produce seeds. In this way, certain desirable characteristics that the parents have in common will probably be perpetuated by the offspring.

Certain variants with unusual characteristics, such as three leaves to a node instead of the usual two leaves, can be inbred continuously until all progeny carry the trait. One problem with inbreeding is that other desirable characteristics may be lost as the new population becomes more homogeneous. Inbreeding plants indoors seems to lead to a loss in potency by the fourth generation. (Preceding generations were considered comparable to the original imported grass.)

Outbreeding is crossing two different varieties. Offspring from parents of two different varieties are called hybrids. *Cannabis* hybrids exhibit a common phenomenon in plants called "hybrid vigor." For reasons not wholly understood, hybrids are often healthier, larger, and more vigorous than either of their parents.

Figure 84. *Upper left:* An old female reversing to male flowering. *Lower left:* Three leaves to a node (trifoliate). *Upper right:* A plant with three leaves to a node alternating with one leaf on next node. *Lower right:* Three-leafed plants sometimes split into two growing shoots.

A reference to cannabinoid content of hybrids from crosses between chemotypes was made in a 1972 study by the Canadian Department of Agriculture: "The ratio of THC to CBD in hybrids was approximately intermediate between the parents . . . there was also occasionally a small but significant deviation toward one of the parents—not necessarily the one with the higher or the lower ratio of THC to CBD."[51] This means that a cross between a midwestern weedy hemp (type III) and a fine Mexican marijuana (type I) would yield offspring with intermediate amounts of THC and CBD, and which hence would be considered type II plants.

Homegrowers have mentioned that inbreeding plants often led to a decrease in potency after several generations. Outbreeding maintained potency, and sometimes (some growers claimed) led to increases in potency.

One area in which breeding can be useful for homegrowers is the breeding of early-maturing plants for northern farmers. Farmers in the north should always plant several varieties of marijuana. Mexican varieties generally are the fastest to mature. Individual plants that mature early and are also satisfactorily potent are used for the seed source in next year's crop. This crop should also mature early. Some growers cross plants from homegrown seed with plants from imported seed each year. This assures a maintenance of high-potency stock.

Potency Changes Over Generations

It is well-established that plants of the P_1 generation (parentals, or the first homegrown plants from imported seed) maintain their chemical characteristics. (For example, type I plants yield type I progeny whose cannabinoids are about equal both quantitatively and qualitatively to those in their native grown parents.) This fact is shown by Table 25.

In the study[66] from which Table 25 has been adapted, individual plants within varieties differed by more than four times in CBD content and by more than three times in THC content. The researchers also noted that illicit marijuana samples contained proportionately less leaf material and proportionately more stem and seed material than samples grown in Mississippi. (Mississippi samples may be more dilute.) New Hampshire and Panama samples were nearly equal in terms of the sum of THC plus CBN.

Table 25.

Cannabinoid Content versus Place Grown[66]

Variant	Place Grown	Year	Average Percentage by Weight[a] of		
			CBD	Δ^9-THC	CBN
Thailand	Thailand	1969	0.14	4.8	0.12
	Mississippi	1970	0.17	3.4	0.11
Minnesota	Minnesota	1968	1.7	0.038	0.15
	Mississippi	1969	1.6	0.077	0.14
Panama	Panama	1969	0.29	3.2	0.80
	New Hampshire	1970	0.38	4.0	0.01

[a]Weight of cleaned, dried flowering tops.

One of the questions that persists in marijuana lore is what effect if any a change in latitude has on the plant chemotype over a period of generations. Nondrug types of *Cannabis* usually originate above 30 degrees latitude in temperate areas. Drug types of *Cannabis* usually originate in tropical or semitropical areas below the 30-degree parallel. Whether this is due entirely to cultural practices is questionable. More likely, the environment (natural selection) is the prime force, and cultural practices reinforce rather than determine chemotype.

Cannabis is notorious for its adaptability. Historically, there are many statements that the drug type of *Cannabis* will revert to the "fiber" type when planted in temperate areas, whereas the fiber type will revert to the drug type after several generations in a tropical area. That a change in chemotype is actually caused by transfer between tropical and temperate areas has not been verified scientifically. (Such studies are ongoing in Europe.) If such changes occur, it is also not known whether the change is quantitative (the plant produces less total cannabinoids) or whether it is qualitative (succeeding generations, for example, change from being high in THC and low in CBD to being high in CBD and low in THC).

We believe that qualitative changes can occur within a few generations, but can only guess what environmental factor(s) might be responsible for such a change. Probably the change has more to do with adaptation of general growth and developmental characteristics than with particular advantages that production of either CBD or THC may bestow upon the plants.

The reason we suspect adaptations of growth characteristics to be the cause of a change in chemotype is that these changes occur rapidly in evolutionary terms, in a matter of several generations. This rapidity implies that some very strong selective pressures are acting on the plant populations. Also, changes in the chemotype seem to occur globally, which implies that the selective pressures responsible are globally uniform rather than local phenomena. Such globally uniform pressures might be light intensity, daylength, ambient temperatures, and the length of the growing season. For example, in populations adapting to temperate areas, those plants that are able to grow well under relatively lower light intensity and cooler temperatures, and which are able to complete development in a relatively short growing season, would be favored over siblings with more tropical characteristics.

Adaptation acts on populations by means of whole organisms which are reacting to a total environment. Shifts in the chemotype of the population are probably linked genetically to the strong selective pressures exerted on the populations by the need to adapt general growth and maturation to either northern (temperate) or southern (tropical) conditions.

CUTTINGS

Marijuana growing often transcends the usual relationship between plant and growers. You may find yourself particularly attached to one of your plants. Cuttings offer you a way to continue the relationship long beyond the normal lifespan of one plant.

To take a cutting, use scissors or a knife to clip an active shoot about four to six inches below the tip. *Cannabis* does not root easily compared to other soft-stemmed plants. Cuttings can be rooted directly in vermiculite, Jiffy-Mix, a light soil, or in a glass of water. The cutting is ready to plant when roots are about an inch long, in about three to four weeks. A transplant compound such as Rootone can be used to encourage root growth and prevent fungi from forming.

Cuttings sometimes grow in surprising ways. One way is referred to as dwarfed growth or "bonsai marijuana." The plant develops short internodes and stout, short branches, but normal-sized leaves. Dwarf growth often occurs when cuttings are taken in an early growth stage. Usually plants will revert to normal growth after about a month or two of dwarfed growth (see Plate 15). Cuttings taken during flowering continue to grow and flower normally once they root.

Another phenomenon seen with cuttings is rejuvenation. The new leaves that form have progressively fewer blades until there is only one blade per leaf, much like the first, true leaves of the seedling. New leaves then follow the normal progression, increasing in size and number of blades. It is as though taking a cutting disrupts the plant's normal growth sequence (genetic program), so that the plant goes back to the beginning and starts over. Rejuvenations can also be induced on whole plants by application of continuous light.[106]

A group of organisms that are genetically identical is called a clone. A group of cuttings taken from the same plant constitute a clone. (Generally, each cell in any organism inherits a replica set of genes aligned on chromosomes which are identical to the original set inherited at conception. To put the situation in simplified terms, cells are specialized in form and function primarily because different genes are either "off" or "on" in each different type of cell.) Clones eliminate genetic differences between individuals, and hence are particularly useful in scientific experiments. By using clones, one can attribute variations between individuals specifically to outside factors. This would be desirable when testing, for example, the affect of fertilizers on potency. Unfortunately, marijuana scientists have not yet taken advantage of this otherwise commonly used research tool.

GRAFTING

One of the most persistent myths in marijuana lore concerns grafting *Cannabis* to its closest relative *Humulus*, the hops plant of beer-making fame. The myth is that a hops scion (shoot or top portion of the stem) grafted to a marijuana stock (lower stem and root) will contain the active ingredients of marijuana. The beauty of such a graft is that it would be difficult to identify as marijuana and, possibly, the plant would not be covered under the marijuana statutes. Unfortunately, the myth is false. It is possible to successfully graft *Cannabis* with *Humulus*, but the hops portion will not contain any cannabinoids.

In 1975, the research team of Crombie and Crombie grafted hops scions on *Cannabis* stocks from both hemp and marijuana (Thailand) plants.[205] *Cannabis* scions were also grafted to hops stocks. In both cases, the *Cannabis* portion of the graft continued to produce its characteristic amounts of cannabinoids when compared to ungrafted controls, but the hops portions of the grafts contained no cannabinoids. This experiment was well-designed and carried out. Sophisticated methods were used for detecting THC, THCV, CBD, CBC, CBN, and CBG. Yet none of these were detected in the hops portions.

The grafting myth grew out of work by H.E. Warmke, which was carried out for the government during the early 1940's in an attempt to develop hemp strains that would not contain the "undesirable" drug.[58] The testing procedure for the active ingredients was crude. Small animals, such as the water flea *Daphnia*, were immersed in water with various concentrations of acetone extracts from hemp. The strength of the drug was estimated by the number of animals killed in a given period of time. As stated by Warmke, "The *Daphnia* assay is not specific for the marijuana drug . . . one measures any and all toxic substances in hemp (or hop) leaves that are extracted with acetone, whether or not these have specific marijuana activity." Clearly it was other compounds, not cannabinoids, that were detected in these grafting experiments.

Unfortunately, this myth has caused some growers to waste a lot of time and effort in raising a worthless stash of hops leaves. It has also led growers to some false conclusions about the plant. For instance, if the hops scion contains cannabinoids, the reasonable assumption is that the cannabinoids are being produced in the *Cannabis* part and translocated to the hops scion, or that the *Cannabis* root or stem is responsible for producing the cannabinoid precursors.

From this assumption, growers also get the idea that the resin is flowing in the plant. The myth has bolstered the ideas that cutting, splitting, or bending the stem will send the resin up the plant or prevent the resin from going down the plant. As explained in our discussion of resin glands in Chapter 2, these ideas are erroneous. Only a small percentage of the cannabinoids are present in the internal tissues (laticiferous cells) of the plant. Almost all the cannabinoids are contained and manufactured in the resin glands, which cover the outer surfaces of the above-ground plant parts. Cannabinoids remain in the resin glands and are not translocated to other plant parts.

We have heard several claims that leaves from hops grafted on marijuana were psychoactive. Only one such case claimed to be first hand, and we never did see or smoke the material. We doubt these claims. Hops plants do have resin glands similar to those on marijuana, and many of the substances that make up the resin are common to both plants. But of the several species and many varieties of hops tested with modern techniques for detecting cannabinoids, no cannabinoids have ever been detected.[212]

The commercially valuable component of hops is lupulin, a mildly psychoactive substance used to make beer. To our knowledge, no other known psychoactive substance has been isolated from hops. But since these grafting claims persist, perhaps pot-heads should take a closer look at the hops plant.

Most growers who have tried grafting *Cannabis* and *Humulus* are unsuccessful. Compared to many plants, *Cannabis* does not take to grafts easily. Most of the standard grafting techniques you've probably seen for grafting *Cannabis* simply don't work. For example, at the University of Mississippi, researchers failed to get one successful graft from the sixty that were attempted between *Cannabis* and *Humulus*. A method that works about 40 percent of the time is as follows. (Adapted from [205].)

Start the hops plants one to two weeks before the marijuana plants. Plant the seeds within six inches of each other or start them in separate six-inch pots. The plants are ready to graft when the seedlings are strong (about five and four weeks respectively) but their stem has not lost their soft texture. Make a diagonal incision about halfway through each stem at approximately the same levels (hops is a vine). Insert the cut portions into each other. Seal the graft with cellulose tape, wound string, or other standard grafting materials. In about two weeks, the graft will have taken. Then cut away the unwanted *Cannabis* top and the hops bottom to complete the graft. Good luck, but don't expect to get high from the hops leaves.

POLYPLOIDS

H.R. Warmke also experimented with breeding programs during the war years. Polyploid *Cannabis* plants were produced by treatment with the alkaloid colchicine. Colchicine interferes with normal mitosis, the process in which cells are replicated. During replication, the normal doubling of chromosomes occurs, but colchicine prevents normal separation of the chromosomes into two cells. The cell then is left with twice (or more than) the normal chromosome count.

Warmke's experiments concluded that polyploids contained higher concentrations of the "active ingredient." However, the procedure for measuring that ingredient was much the same as described for grafting, with probably similar shortcomings.

Polyploid *Cannabis* has been found to be larger, with larger leaves and flowers. Recent experience has shown that polyploids are not necessarily higher in potency. Usually they are about equal to diploid siblings.

Colchicine is a highly poisonous substance. The simplest and safest way to induce polyploids is to soak seeds in a solution of colchicine derived from bulbs of winter or autumn crocus (Colchicum). Mash the bulbs and add an equal part of water. Strain through filter paper (or paper towels). Soak seeds in the solution and plant when they start to germinate. Cultivate as usual.

Only some of the seeds will become polyploid. Polyploid sprouts generally have thicker stems, and the leaves are often unusually shaped, with uneven-sized blades. Leaves also may contain more than the usual number of blades. As the plant grows, leaves should return to normal form, but continue to be larger and with more blades.

If no polyploids sprout, use less water in preparing the solution.

Colchicine is also a prescribed drug for treatment of gout and is taken in pill form. These usually contain .6 mg per tablet. Use 10 tablets per ounce of water, and soak the seeds as described above.

Colchicine is also sold by mail-order firms which advertise in magazines such as *Head* or *High Times.*

Because colchicine is a poison, it should be handled carefully. It is not known if plants from seeds treated with colchicine will contain a harmful amount of colchicine when the plants are grown. Harm is unlikely, because the uptake by the seed is so small, and because the colchicine would be further diluted during growth, as well as diminished by smoking. But we cannot guarantee that you can safely smoke colchicine-treated plants.

Chapter Nineteen
Effects of the Environment on Potency

This chapter deals only with the influence of the environment on the potency of your crop. Differences or changes in potency can also result from inherent differences between plants, such as in their variety or growth stage, from chemical degradation of the harvested marijuana, and from genetic processes that take place over several generations of plantings.*

We have emphasized that heredity is the most important factor that determines potency. Potent marijuana grows from seeds of potent marijuana. A healthy, mature plant bears an abundance of flowers, guaranteeing you a potent harvest.

Some researchers have investigated the impact of the environment on relative potency, since this question is of interest to officials concerned with marijuana control as well as to marijuana growers. Their primary goal has been to discover the gross effects of different environments rather than to single out the effects of any particular factor. A consensus is that the impact of environment on potency is small relative to that of the plant's heredity. Nevertheless, where scientists have commented on this question, the common denominator for higher potency has been stress.

· STRESS

Stress is anything that detracts from the plant's health or vigor. Environmental factors such as competition from other plants, low water availability, and poor soil conditions are examples of stress factors.

In many marijuana-growing cultures, farmers have practices

---

*For discussions of these other causes of differences in potency, see the following sections: "*Cannabis* Chemotypes" in Chapter 2; "Inherent Variations" in Chapter 3; "Breeding" in Chapter 18; "Potency and Decomposition" in Chapter 20; and "Storage" in Chapter 21.

that are stress-related; splitting the base, severe pruning, bending or contorting the stem, and water deprivation are common examples. Of course, the fact that marijuana-growing cultures have such practices does not mean that these practices actually increase potency, or that this is, or ever was, their intent. Their original meaning may well have been forgotten centuries ago. For instance, cultivation of sinsemilla has been practiced for centuries, not for potency, but because the seedless product is easier to process or smoke.

There does seem to be some underlying relationship between stress and higher potency. Stress factors may slow growth in general, but at the same time, may not slow the synthesis of cannabinoids. Potency may be affected in much the same way by growth factors that are not considered stressful. As described previously, marijuana plants grow more compactly and have smaller leaves under conditions of relatively warm temperatures, or strong sunlight, or a dry atmosphere; they grow taller and have larger leaves when grown under cool temperatures, moderate light, or a humid atmosphere. Higher relative potency seems to correlate with conditions which favor compact development of the plant and its parts.

The rate of cannabinoid synthesis relative to photosynthesis may be affected in ways not apparent. Sunlight, for instance, is a growth factor. In almost all cases, the more sunlight the plants receive, the faster and larger they will grow. Yet plants grown with intense sunlight seem to maximize potency. Intense sunlight can raise plants' internal temperatures to levels that interfere with the photosynthesis cycle. Absorption of light energy and conversion to biochemical energy continues unimpeded, but the synthesis of sugars is impeded. (Under a midday sun, this phenomenon has been observed in other field crops.[206]) In marijuana, cannabinoid synthesis may continue unaffected at these higher temperatures. This might account, in part, for the slightly higher potency of plants grown in tropical zones.

The subject of potency is mired in confusion and mystery, largely because of fertilizer and soil ads, marijuana-growing books, and individuals who promise ways of increasing potency or growing super grass. There are no magic formulas or secrets to divulge that will make or break the potency of your crop. We have tried to play down this type of thinking throughout this book. Choice of seed, and a harvest of well-developed buds, far outweigh any other factors in determining potency.

We know of no one who has demonstrated that manipulation of any particular environmental factor leads to higher potency. This lack of demonstration probably exists for two reasons: (1) environmental effects on potency are relatively small compared to the effects of inheritable traits, and hence are not easy to discern; and (2) "increased potency" is difficult to prove.* The variations in potency within any variety, and within each plant, require stringent methods of sampling for comparative tests. And since potency also changes with time, meaningful comparisons can be difficult to make. Scientific papers reflect this difficulty.

An experiment on potency must account for inherent variations in potency before environmental effects can be analyzed. Samples would need to be equivalent in terms of variety, growth stage and development, sex, plant part, and the position of the part on the plant. A simple way to do this would be to harvest females when each reached full bloom and then compare the uppermost buds from each plant.

Most of the research on potency done to date either has not reported sampling techniques or did not account for certain inherent variations. In the extreme case, all vegetative matter from one plant was mixed together, and the THC concentration in a sample of this matter was compared with that in a similar sample of mixed marijuana from another plant. Such practices can give misleading results. Consider the fact that an unhealthy plant will have dropped many of its lower, less-potent leaves. A healthy plant has more leaf overall and retains more of its lower leaves. There is a good chance that the unhealthy plant will test higher in average THC content, because proportionately more of the sample from it consists of the uppermost leaves and shoots, which are relatively more potent. Such a sampling error could create the impression that stress is positively correlated with potency.

Marijuana scientists have recognized the need for testing equivalent samples and for setting standards for testing. In 1974,

---

*Most scientific experiments are observations, and test effects or compare results rather than try to prove something.

Figure 85. *Left:* One way to stress the plants is to strip the leaves (Mexican). *Right:* Colas develop more thickly when some leaves are left on the plant.

the English scientist John Fairburn[68] published a number of well-controlled experiments concerning potency and light. This is a hopeful sign that more meaningful experiments will be forthcoming.

We have said that the common denominator for increasing potency is stress. Let's put that in perspective.

You cannot go wrong if you grow the largest and healthiest plants possible. *Our experience has been that the most potent plants are more often the healthiest and most vigorous in the garden.*

Factors that limit growth rate are probably related to potency, and if growth rate is relatively slower, cannabinoid concentrations may be higher. Plants whose average yield is six ounces may be slightly more potent than plants whose average yield is eight ounces. Factors related to potency affect growth rate rather than ruin the plant's health. When a plant is so traumatized that it is barely surviving, potency as well as growth rate declines.

Obviously, if the growth rate is slower, the harvest will be smaller. Any difference in potency due to stress is quite small, but the difference in yield can make the difference between harvesting an ounce and harvesting several pounds (i.e., don't get carried away with the practice of stressing the plants). If you wish to stress the plant, wait until it is firmly established and growing well.

Outdoors, don't stress the plant until at least the middle of its life. You want the plant to be large enough to bear a good harvest of buds. Water deprivation is a good method of limiting growth outdoors. However, wilted plants must be watered, or they will die.

Competition from other weeds has been correlated with higher potency in two recent studies.[71,74] You might prefer not to weed your patch after the seedling stage. (Initially, weeding is necessary because indigenous weeds generally outgrow marijuana seedlings.)

Another safe way of applying stress is to remove all large leaves from the plant once it has begun to flower.

Indoors, the plants are already in a delicate state. We advise indoor growers to grow the largest and healthiest plants possible for best results.

NUTRIENTS

Most growers show a keen interest in fertilizing, since it is one factor over which they have some control. Most growers also feel that nutrients, which play such an important part in plant growth, probably have a relationship to the potency, and this is a reasonable assumption. In marijuana lore, potency is sometimes attributed to particular soil types (for instance, red dirt, which is iron-rich) or to presence or lack of certain nutrients (for example, nitrogen or potassium deficiency).

The relationship of potency to soil conditions, in particular the nutrient content, has been looked at recently by several research groups. In two such studies,[71,74] the cannabinoid content of naturalized weedy hemp stands in the Midwest was examined. Variations in potency were then correlated to soil properties, such as N, P, and K content. The two papers came to similar conclusions. First, stands growing in areas where they were under stress tended to produce less biomass (yield) but were more potent overall. Second, when nutrients or other growth factors, such as height and weight of plants or root size, were correlated with potency, potency was almost always correlated positively with positive growth factors. That is, higher potency occurred when the plants were growing with adequate or high amounts of nutrients present, not when nutrients were inadequate.

What appears to be a contradiction (stress leads to higher potency, or good growing conditions lead to higher potency) may be explainable in terms of what these experiments actually measured.

A basic assumption in these studies was that all the plants were relatively homogeneous genetically, since they may have originated from a single stock of hemp grown during World War II. Assuming this is true, then variations between stands would be due to differences in local environmental factors. However, since environmental conditions differed locally for separate stands, one cannot tell whether variations in potency between stands are due to present environmental factors (phenotypic responses) or reflect thirty years of adaptation by each stand to its local environment (genotypic shifts).

It may be that positive growth factors are associated with higher potency in phenotypes (plants now growing), whereas stress leads to higher potency in succeeding generations, because of selective pressure. It is interesting that both papers reported strong positive correlations between higher potency and competition from other weeds, since competition between plants does exert strong selective pressures.

The following list of possible effects of nutrients on potency has been adapted from these four studies.[63, 71, 74, 231]

Nitrogen

Nitrogen was positively correlated with higher potency. One controversy in marijuana lore is whether a nitrogen deficiency during flowering increases potency. We have grown plants with N deficiencies, and they seemed no more potent than those grown with high amounts of nitrogen available. However, the N-deficient plants did produce a much smaller harvest.

Phosphorus

P has been correlated positively with higher potency in all studies that have examined this factor. Phosphorus is necessary for good flower development and seed production. Give the plants a steady supply of phosphorus throughout growth and in particular during flowering.

Potassium

K has been correlated both positively and negatively with potency. More often, it has been found to be negatively correlated. As discussed previously, plants that show some potassium deficiencies may grow well; so you may choose not to treat minor symptoms of K deficiencies. *Cannabis* requires adequate amounts of potassium during all stages of growth. *Cannabis* that shows symptoms of K deficiency often grows vigorously with little harm other than the spotting and the loss of some lower leaves. It should not be necessary to fertilize with potassium during flowering unless deficiency symptoms are severe and the plant has ceased growing.

Calcium

Abundant Ca levels have been consistently correlated with higher potency.

Magnesium

Mg has been negatively correlated with potency. However, this may have been due to the interaction of Ca and Mg, and may reflect Ca's strong positive correlation to potency, rather than the negative effects of Mg per se. Plants that show Mg deficiencies must be fertilized, or they will quickly lose most of their leaves and barely remain alive.

PART 5.
Harvesting, Curing, and Drying

Chapter Twenty
Harvesting

Figure 30 is a hypothetical plot of the increase in potency of a male plant and a female during the course of their growth. (Potency is measured by the percentage by weight of THC in a dried sample of the uppermost leaves or growing shoots until flowers appear.) It shows that generally potency increases as the plant develops. Cues such as phyllotaxy changes and rate of growth are helpful indicators to changes in development and the chronological age of the plant has little significance.

The development of the cannabinoids, resin glands, and, in practical terms, the potency in the living plant is not clearly understood. We believe that, for the most part, potency does not increase steadily throughout the entire plant. Rather, each plant part reaches a point of maximum potency as it individually develops. A leaf that is formed when the plant is four weeks old does not increase in potency during the rest of the season. To say that potency is increasing means that the leaves that are now forming are more potent than those previously formed.

We also believe that cannabinoid formation is very fast as each plant part forms. Once matured (for example, when a leaf is fully expanded), cannabinoid synthesis probably continues, but at a very reduced rate, which may actually be exceeded by the rate at which the cannabinoids are decomposing. This is one reason why the potency can decrease as well as increase during growth, especially late in the season, after the flowers have formed. The practical aspects of these points are detailed in the following sections.

HARVESTING DURING GROWTH:
LEAVES AND GROWING SHOOTS

Leaves

We have emphasized that you should harvest grass during the course of the season. One reason is to assure yourself a return for your efforts. It is a sad commentary on our times that the greatest danger in growing marijuana outdoors is that the plants may be ripped off. On a more positive note, vegetative shoots and leaves can be surprisingly potent and should be sampled.

The potency of each new set of leaves is higher than the last pair until a plateau is reached, usually during the middle of vegetative growth. Thereafter potency of new leaves stays about the same as in those preceding. Often there is a noticeable decline in potency just prior to flowering. Leaves that form during flowering are usually more potent than those formed during the vegetative plateau. Leaves that form after the bloom are less potent.

Of course, not all varieties or individual plants will follow this rhythm. Faster-developing plants may reach the plateau sooner, and slower plants later. Potency of plants that have a longer life cycle may stay at the vegetative plateau for several months. Some plants do not seem to experience any drop in potency before flowering. Potency of these plants continues to increase gradually after the initial quick increase during early vegetative growth.

Whenever you harvest green leaves during growth, you should always take the uppermost leaves, since these are the most potent. Also, the smaller leaves that form on the branches are more potent than the large leaves on the main stem. These large stem leaves (fan, shade, or sun leaves) are often the first leaves that growers pick. But these are the least potent of all the leaves, and they may not get you high at all. As long as these leaves are healthy and green, let them stay on the plant for the plant's growth. Many growers simply use these leaves for mulch or compost as they die.

Don't think that you should harvest each leaf as soon as it appears; this procedure would seriously affect normal growth and result in a small harvest of buds. The potency of individual leaves does not increase during the course of the season, but the decrease in potency is not great. Some of the loss in potency may even be made up for by the loss in tissue weight that a leaf experiences as it dies. Many growers prefer to harvest leaves during growth only after they lose color, preferring the taste of the smoke to that of green leaves.

Leaves should always be harvested if they die; with indoor gardens, remove any leaves that show signs of insects or other pests.

Do keep yourself supplied with grass (that is the reason you are growing the plants); just don't overdo it. The main harvest is made up of buds, and you want a large, healthy plant that can support vigorous flowering. The larger and healthier a plant is, the more leaf you can harvest without seriously affecting the plant.

Growing Shoots

You may prefer not to clip the growing shoot of the main stem. This forms the largest and most potent cola by harvest time. Plants grown close together usually are not clipped, so that the plants may grow as tall as possible. Where there is much space between plants, the main shoot is clipped to encourage the plant to develop its branches, which fill the available space.

The potency of growing shoots follows the rhythm described for new leaves. However, growing shoots can be the most potent parts of the harvest when picked at the right time. Shoots sometimes reach a very high peak of potency during the middle of vegetative growth. Outdoor gardens should be sampled from mid-June through July, since this is the period in which the shoots usually reach their peak.

Potency also fluctuates according to local weather conditions. Try to harvest after a period of clear, sunny weather. Potency may decline for several days after a period of cloudy weather or heavy rainfall. After a heavy rain, harvest the shoots a week or two later, since shoots often peak in potency during a burst of fast growth.

Growing shoots can be harvested from each plant at least twice during growth. The first clipping may not give you much worthwhile grass, but it is done when the plants are young (roughly six weeks old) to force the plant to develop several growing shoots which are harvested about six to eight weeks later. The main shoot is clipped, leaving about four or five nodes below the cut. Two shoots should start to grow from each node, the strongest at the top of the plant and the weakest at the bottom. (This difference is more pronounced under artificial light, since the light is strongest on the top of the plants.) Each plant should produce at least six strong growing shoots after this first clipping. The yield from growing shoots can be considerable (especially during the summer marijuana drought) and will probably keep you supplied until the main harvest.

A third harvest of shoots can be made later if the plants have a long growing season or are indoors. You don't want to clip shoots from the plant just prior to or during flowering, since doing so cuts down on the harvest of buds. Each plant should have at least twelve growing shoots after being clipped twice previously. You might harvest only a few shoots from each plant if the time for flowering is near.

MALE PLANTS

Male plants usually do not have the dramatic increase in potency during flowering that the females do. Male flowers take about two weeks to mature, from the time they are first visible as tiny knob-like buds. New flowers continue to appear for several weeks.

When male flowers open and are about to release pollen, they reach their maximum potency. Since all flowers do not mature at the same time, for maximum potency the plants should be harvested after the first few flowers have opened.

Male flowers actually make up little of the total weight of the harvest, and few new leaves form once flowering begins. There is no significant loss in either potency or yield if the male is harvested before its flowers open. Once male flowers appear, there is little change in their potency. Also, once the flowers do open and release pollen, they shortly fall from the plant and are lost to the harvest.

Males should therefore be harvested before any flowers open unless you want the females to produce seeds. In a small garden, male flower clusters can be individually harvested as they mature. Most growers treat male flowers more as a novelty. Potency of male flowers is quite variable, and seldom are they as good as the female flowers. To remove male plants, cut them near the base of the stem. Don't rip them up by the roots if they are near females that will be left to grow.

Male plants normally begin to lose their vigor after the initial bloom. When the weather is mild, or the plants are indoors, they can be encouraged to bloom a second and sometimes a third time before they finally die.

HARVESTING FEMALE BUDS

The decision of when to harvest females can be simplified by understanding that you want to pick the buds after they have developed fully, but before degradation processes begin to lower potency. There are two criteria you can use to tell when the plants have reached full bloom. The first is recognizing the rhythm with which the plants are blooming. A second is the condition of the flowers as judged by the health of the stigmas and the color of the resin.

Sinsemilla

Since sinsemilla flowers are not pollinated, the flowering period may last for many weeks. The most common rhythm for sinsemilla is that plants go through a stage of rapid bud formation, and the plants do indeed bloom. This bloom often lasts four to five weeks. The bloom ends when the rate at which new flowers form noticeably declines. At this time you should be able to sense that the bloom is completed. Buds are at their peak potency about one week after flower formation slows. This is the time to harvest. True, the plant may continue to grow slowly, but the main harvest is ready and should be taken.

With sinsemilla, some marijuana varieties have an extended bloom that may last more than two months. With this rhythm, the rate at which the buds form is drawn out, and progresses at a slower but steadier pace. The point at which the bloom is essentially over may not be as obvious as in the first case. Here, use the condition of the buds to make your decision. Stigmas wither first at the base of the buds (older flowers). Those stigmas at the top of the buds (younger) will still be white and healthy, although their tips are often brown. Harvest the plants

Figure 86. *Left:* Sinsemilla buds at harvest. *Right:* A sinsemilla cola. Also see front cover for sinsemilla at harvest.

when about half the stigmas in the buds have withered. The coating of resin glands should still be clear or white, with only a few golden or browned gland heads.

A third type of flowering rhythm is sometimes seen on plants from Thai seed. Flower buds bloom and ripen at different times. These plants also have an extended flowering stage that can last for over two months. You may choose to harvest individual buds, colas, or branches as they ripen.

Seeds

If your primary interest is seeds, the plant should be harvested after the seeds have developed their mature color. Mature seeds can be seen splitting their sheaths or bracts. When enough seeds have ripened, the plants should be harvested. If the plants are left in the ground and die, many of the seeds will fall from the plant.

For most growers, potency will be of primary interest, seeds only secondary. With seeded marijuana, flowering is initially rapid until the plant is well-pollinated. If pollination occurs early in flowering, the plants often bloom for another week or two. Generally, you want the plants to flower for at least four weeks before picking, and usually longer, about six to seven weeks.

With seeded marijuana, the bloom is of shorter duration than with sinsemilla. Once growth slows, wait another two to three weeks before harvesting. All the seeds may not be matured, particularly at the top of the bud. But potency of the buds should be about maximum at this time.

Figure 87. Partially seeded marijuana ready for harvest (Colombian).

WEATHER

Because of such variables as variety and growing conditions, there can be so much variation in the ripening process that no one criterion for judging when maximum potency is reached will be reliable for all cases.

Warm, sunny weather encourages rapid flowering and a long period of receptivity by the stigmas. Cool, rainy weather can wither the stigmas and dampen the vigor of the bloom.

If a brief frost or long, cool rain has withered the stigmas, use the plants' growth as a guideline, because ultimately this is the most important criterion. You want the buds to reach a mature size, and to ripen for about another week. You do not want the buds to be left on the plant longer than necessary.

Ideally, harvesting should follow a period of warm, sunny weather. In northern and mountainous parts of the country, many tropical varieties will not flower until late in the season, when the weather has cooled and nighttime frosts are threatening. Most mature plants can withstand mild frosts and continue to grow well if daytime temperatures are mild. In this case, let the plants mature, since formation of the buds is more important than the weather in determining potency. Watch the plants carefully, and harvest when the buds reach mature size. Marijuana killed by frost may smoke harshly, but potency does not seem affected. Well-formed buds should be picked if heavy rains are expected. Cannabinoids are not water-soluble, but gland heads will be washed away.

Barring a catastrophe, such as a long frost, death to *Cannabis* is usually not sudden. The plants will continue to grow, and may in fact rejuvenate the next year if the stalks are left in the ground. But after the main bloom, the growth that follows is usually much less vigorous and sometimes forms abnormally. Leaves at this time are simplified, and have one blade. Later leaves are smaller, and tend to have entire margins (no serrations). Sometimes they are twisted or misshaped, as are the flowers that form along with them. This slow growth that follows the initial bloom will contribute little to the weight of the harvest. Additionally, this post-bloom growth is much less potent than the original bloom. Resin glands on these plant parts are few and poorly developed. When this abnormal growth forms, the time for harvesting is past. See Figure 83.

When a plant seems to persist in growing, and you are not sure the bloom is past, the best procedure to follow is to try for a double harvest.

DOUBLE HARVESTS

Most marijuana plants take at least five months to reach maturity. Once the plant has reached maturity, it is forming its most potent marijuana, and should not be cut down completely.

Figure 88. Buds beginning to form for a double harvest under lights. First crop was harvested four weeks earlier.

You can often induce the females to flower a second (and sometimes a third) time, especially if the plants are indoors or if the weather is expected to stay mild for several more weeks.

To encourage a second bloom, first take the bulk of the harvest: all but the smallest buds, and most of the leaf. Some green leaves should be left on the plant to maintain the plant's growth. After harvesting, give the plants a thorough watering, and water with a soluble, complete fertilizer that provides a good supply of both N and P. This will encourage new growth and continued flowering.

Indoors, the best procedure is to treat the plants like a hedge. Cut all the plants back to equal heights, about two to three feet tall. Remove most of the grass, but again leave a few green leaves on the plant. Don't remove lower branches even if they are leafless, since these will sprout again. Lower the light system to the tops of the plants, and maintain the daily cycle at about 12 hours. The second crop of buds will be ready for harvest in four to eight weeks. With this system, the plants appear like dense hedges of buds. If the second crop of buds forms quickly, you should try for a third crop. Continue to fertilize the plants regularly, and watch for signs of magnesium deficiencies, which often show up when the plants have been growing for an extended time.

Double and triple harvests are one of the benefits of indoor growing. Although plants are relatively small indoors, the original harvest of buds can be tripled in the next four months.

POTENCY AND DECOMPOSITION

We have said that when buds are picked too late, the potency may decline because of decomposition of the cannabinoids, especially THC.

In Chapter 21, Tables 26-29 give measured rates of decomposition of the major cannabinoids due to exposure to light and air. Light rapidly decomposes THC into unknown products (possibly polymers[122,164]). Light also converts CBD to CBS and CBC to CBL. Air (oxygen) slowly converts THC to the less active CBN. Conversion to CBN is hastened by higher temperatures.

Degradative processes do not occur as quickly in the living plant as when the cannabinoids are purified or in solution, as is shown by the data in Tables 27-30 in Chapter 21. Resin glands seem to function well in storing the cannabinoids in dried plant material. However, the rates of decomposition in Tables 27 and 28 are for samples exposed to north light and a maximum of 80° temperatures. Temperature would be higher, and light stronger, under full sunlight.

Studies with fresh plant material usually show negligible CBN content in fresh marijuana from immature plants. When mature buds are tested, their CBN content is generally equal to at least five percent of their THC content. When growing temperatures are higher, such as in the tropics. CBN content can account for more than 20 percent of the original THC. Even if we assume a low figure, such as five percent conversion of THC to CBN, there is actually a much greater decline in THC content because of the simultaneous degradation of THC by light.

When the slow rate at which THC oxidizes to CBN is considered, five percent decomposition in a period of less than two months represents considerable exposure of the THC to air, and most of this exposure occurs in the last critical weeks when the resin glands begin to degenerate. Plates 8 and 11-13 show the condition of the resin glands on several different kinds of marijuana.

Stalked glands that cover the female flower bracts sometimes rupture or secrete cannabinoids through pores in the gland head. Secretion is not a continuous flow, but more of an emptying of the glands' contents. At this time, gland heads may dehisce. Also, because of their abundance and raised positions, resin glands on the female bracts are exposed to strong sunlight and possible physical damage. These conditions may explain the significant decline in potency of buds that are overripe.

Leaves are also affected by decomposition of the cannabinoids, but not as quickly or seriously as the buds, probably because the resin glands on the leaves are most numerous on the undersurface, where they are somewhat protected from light. These glands rarely rupture or secrete cannabinoids. Often they are intact, clear, and apparently unchanged for many weeks on the living plant.

As the plates show, one can, with the naked eye, see the glands change color, from colorless or white to golden, and then to reddish or brown. THC is colorless. If the color changes of the resin do indicate decomposition of THC, then decomposition in the stalked glands that cover the buds can be considerable.

We have smoked buds that seemed to lose about half their potency when left on the plant for an additional three weeks. Color changes are after the fact. If many of the glands are beginning to brown, the grass should be harvested.

TIMING THE HARVEST

Many growers will disagree with us on when the best time is to harvest the buds (female plants). When the plants are left in the ground, and are alive but past the main bloom, the resinous qualities of the plant may become more apparent. The bracts and tiny leaves may swell in size, and the leaves feel thicker. The coating of resin glands will change color. Leaves often yellow and fall from the plant. Much of the green color in the flowering buds may also be lost. Harvests of these buds more closely resembles commercial Colombian grass than typical homegrown. The resin content of the dried buds may be higher, and the grass will smoke more harshly than if the buds were younger when picked. You may prefer these qualities in your grass, and some growers insist this grass is stonier. We feel that the grass will give you the highest high when it is picked as described previously. Smoking is a personal experience, and you should try different approaches and come to your own conclusions.

The first time you grow marijuana is largely a learning experience. Most growers can't wait to start their second crop, because they are certain that they'll improve on both the quantity and the quality of their crop, and this is usually true. The wise grower will not put all his proverbial eggs in one basket. It is a good idea to monitor potency by taking samples every few days when harvest time is drawing near, just as such monitoring is for deciding when to harvest growing shoots during vegetative growth.

In any garden, some of the plants will mature sooner than others. Use the plant(s) that is earliest to mature to decide at what point in its development the plant reaches maximum potency. This finding then serves as a guide for harvesting the rest of the plants.

Try to use buds from approximately the same position on the plant each time you sample. Take only enough to make a joint or two. The more you standardize your testing (and this includes your smoking evaluation), the more accurate your results may be.

FINAL HARVEST

The time of harvest is a time of joy. It is also a time for caution. Unless the safety of your garden is assured, you will want to harvest quickly, quietly, and as efficiently as possible. Ideally, each plant is harvested as it matures, but some of you will have to harvest all at once.

It is best to take cardboard boxes or large, sturdy bags to carry the harvest. You want to harvest the plants with as little crushing or damage to the flowers as possible.

Bring a strong knife, heavy shears, or clippers for cutting the stalks. The quickest way to harvest is to cut each plant at its base. Once the plants are on the ground, cut the stalks into manageable lengths for boxing or bagging. Separate large branches as needed for packing.

The bagged or boxed material should be moved to the curing or drying area as soon as possible. If you let the plants sit in the trunk of a car or in plastic bags, they will start to ferment and smell in less than a day.

Chapter Twenty-One

After the Harvest

Once the marijuana plant is harvested, it ceases to produce cannabinoids and resins, and the main changes in potency that occur are degradative. However, when the material is handled carefully, dried or cured properly, and then stored well, little degradation will occur. During drying or curing, the resin content may seem to increase, as the plant's tissues shrink away from their resinous coating.

More than 70 percent of the fresh weight of the plant is water. Drying is done to evaporate most of this water, so that the marijuana will burn evenly and smoke smoothly. Additionally, the cannabinoids in fresh plant material are mostly in their acid forms, which are not psychoactive. The acid cannabinoids decarboxylate (they lose the gas, carbon dioxide) during the drying or curing processes, which convert them to their psychoactive neutral forms. Decarboxylation is complete if the marijuana is actually smoked. For this reason, no special procedures are needed to decarboxylate the marijuana unless it will be eaten. In that case, the recipe should include a period of dry heating. The heat converts the cannabinoids to their psychoactive neutral forms, and also melts the sharp-pointed cystolith hairs that cover leaves, stems and petioles. Cystolith hairs can cause stomach pains if you eat uncooked marijuana or chew on raw marijuana, which we strongly advise you not to do.

Commercial marijuana is usually composed of just the flower tops (colas), which have been stripped, manicured, cured, and dried. Homegrowers often do not cure their crop before drying, and if the smoke is smooth, there is no reason not to dry it directly. But harsh-smoking marijuana can be cured so that the smoke is smoother. Curing has little affect on potency when done properly.

STRIPPING

Stripping, the removal of large leaves, is usually done soon after harvesting. Fan leaves are stripped because they are much less potent than the colas that they cover, and do not cure as well as the prime material. In commercial growing areas, the fan leaves are often stripped in the field and form a green manure. But fan leaves are sometimes quite potent, especially if they are recent growth. The lower leaves are usually weak, but they can be used in cooking or concentrated in an extract.

The easiest time to strip fan leaves is after they have wilted, because they are easier to pull off when they are limp than when they are turgid. Wilting takes place in less than an hour if the plants are in a well-ventilated space. Plants placed in a plastic bag in a cool area may take a day or more to wilt.

Some growers leave the fan leaves on until the plants have dried. After the buds are removed, they strip the remaining leaves by running their (gloved) hands from the base of stems and branches toward their tips. The fan leaves disintegrate into shake.

GRADING AND MANICURING

Grading and manicuring are important steps in preparing fine grass. Grading is done by separating the plants according to variety, sex, and the particular plant parts. This procedure makes the quality of each particular stash uniform, and the quality of the better grades is not diluted. Plant parts are usually graded as follows: main top colas, small side colas, immature buds, leaves accompanying flowers, and fan and stray leaves. This is important, because the differences in potency will be considerable. For instance, the buds on a Colombian home-grown will be top quality, but the lower leaves will be more like a low-grade commercial Mexican.

Manicuring is done to remove the extraneous leaf from the colas. First the large fan leaves are stripped. The exposed colas are then trimmed with scissors to remove the ends of leaves that stick out from the colas. Plants should be manicured and (usually) graded before drying, since dried material crumbles into shake when handled. Also, leaves dry much more quickly than buds, and different plant parts cure at their own rates.

Male flowers are often treated as a novelty by growers, who make individually rolled sticks from them, as follows. Hang the plants upside down; the leaves will wilt and hang down, covering the male flower clusters. Then roll each cluster within its leaves between the palms of your hands, to compress the cluster into a joint-shaped mass. Dry the "stick" in a warm, dark place. Rolling the grass ruptures many of the glands; so dried sticks should be stored carefully until each is used.

When you handle your crop, you may notice a resin build-up on your hands and the tools you are using. This resin can be collected by rubbing and scraping it into a ball. It makes a quality hash that is several times as concentrated as the grass.

Small quantities of hash can be made by rubbing resinous plant parts across a thin, fine-mesh screen. The resin is then scraped off the screen and rolled into finger shapes. Hash can also be made by thrashing fresh plants over a mesh screen inside an enclosed box whose floor is lined with sheet plastic. A box about one yard square is a suitable size. On one side of the box a hole is made large enough for the colas to be shaken by hand. In this way, the resin glands are knocked loose, but are contained by the box and settle on the plastic.

Plastic or paper should be placed beneath the marijuana during manicuring, grading, or drying. Besides fallen grass, resin glands are also caught. With large quantities of grass, a considerable amount of glands and fine shake can be caught and compressed to a hash-like mass.

CURING

Curing is a process employed to naturally enhance the bouquet, flavor, and texture of marijuana. Curing does not lower potency when done correctly, although poor curing methods often result in some loss of THC.

Curing is not an essential procedure, and many growers prefer the "natural" flavor of uncured grass. Sweet sinsemilla buds usually are not cured.

Curing is most successful on plants which have "ripened" and are beginning to lose chlorophyll. It is less successful on growing tips and other vigorous parts which are immature. These parts may only lose some chlorophyll.

Curing proceeds while the leaf is still alive, for until it dries, many of the leaf's life processes continue. Since the leaf's ability to produce sugars is thwarted, it breaks down stored starch to simple sugars, which are used for food. This gives the grass a sweet or earthy aroma and taste. At the same time, many of the complex proteins and pigments, such as chlorophyll, are broken down in enzymatic processes. This changes the color of the leaf from green to various shades of yellow, brown, tan, or red, depending primarily on the variety, but also on growing environment and cure technique. The destruction of chlorophyll eliminates the minty taste that is commonly associated with green homegrown.

There are several methods of curing, most of which were originally designed to cure large quantities of tobacco. Some of them can be modified by the home grower to use for small marijuana harvests as well as large harvests. The methods used to cure marijuana are the air, flue, sweat, sun, and water cures.

Air Curing

Air curing is a technique developed in the United States for curing pipe and cigar tobacco. It was originally done in specially constructed barns made with ventilator slats which could be sealed; a small shed or metal building can easily be adapted for this use. However, this method of curing works only when there is enough material to keep the air saturated with moisture.

Wires are strung across the barn, and the marijuana plants or plant parts are hung from them, using string, wire twists, or the crooks of branches. The plant material should be closely spaced, but there should be enough room between branches (a few inches) so that air circulates freely. The building is kept unventilated until all the material loses some chlorophyll (green color). This loss occurs rapidly during warm sunny weather because heat builds up, which hastens the cure. In wet or overcast weather, the temperature in the chamber will be cooler, and the process will proceed more slowly. If these conditions last for more than a day or two, unwanted mold may grow on the plants. The best way to prevent mold from forming is to raise the temperature to 90°F by using a heater.

After the leaves have lost their deep green and become pale, the ventilators or windows are opened slightly, so that the temperature and humidity are lowered and the curing process is slowed. The process then continues until all traces of chlorophyll are eliminated. The entire process may take six weeks. Then the ventilators are opened, and an exhaust fan installed if necessary, to dry the material to the point that it can be smoked but still is moist, that is, bends rather than crumbles or powders when rubbed between thumb and forefinger.

Flue Curing

Flue curing differs from air curing in that the process is speeded up by using an external source of heat, and the air circulation is more closely regulated. This method can be used with small quantities of material in a small, airtight curing box constructed for the purpose. Large quantities can be hung in a room or barn as described in Air Curing.

A simple way to control the temperature when curing or drying small amounts of marijuana is to place the material to be cured in a watertight box (or a bottle) with ventilation holes on the top. Place the box in a water-filled container, such as a pot, fishtank, or bathtub. The curing box contains air and will float. The water surrounding the box is maintained at the correct temperature by means of a stove or hotplate, fishtank or waterbed heater, or any inexpensive immersible heater. Temperature of the water is monitored.

With the marijuana loosely packed, maintain water temperature at 90 degrees. After several days, the green tissue turns a pale yellow-green or murky color, indicating yellow or brown pigments. Then increase temperature, to about 100 degrees, until all traces of green disappear. Raise the temperature once again, this time to 115 degrees, until a full, ripe color develops. Also increase ventilation at this time, so that the marijuana dries. Plants dried at high temperatures tend to be brittle; so lower the temperature before drying is completed. This last phase of drying can be done at room temperature, out of the water bath. The whole process takes a week or less.

Marijuana cured by this technique turns a deep brown color. Immature material may retain some chlorophyll and have a slight greenish cast. Taste is rich yet mild.

Sweat Curing

Sweat Curing is the technique most widely used in Colombia. Long branches containing colas are layered in piles about 18 inches high and a minimum of two feet square, more often about ten by fifteen feet. Sweat curing actually incorporates the fermenting process. Within a few hours the leaves begin to heat up from the microbial action in the same way that a compost pile ferments. The change in color is very rapid; watch the pile carefully, so that it does not overheat and rot the colas. Each day unpack the piles, and remove the colas that have turned color. Within four or five days, all the colas will have turned color. They are then dried. One way to prevent rot while using this method is to place cotton sheets, rags, or paper towels between each double layer of colas. The towels absorb some of the moisture and slow down the process.

Sweat curing can be modified for use with as little marijuana as two large plants. Pack the marijuana tightly in a heavy paper sack (or several layers of paper bags), and place it in the sun. The light is converted to heat and helps support the sweat.

Another variation of the sweat process occurs when fresh undried marijuana is bricked. The bricks are placed in piles, and they cure while being transported.

A simple procedure for a slow sweat cure is to roll fresh marijuana in plastic bags. Each week, open the bag for about an hour to evaporate some water. In about six weeks, the ammonia smell will dissipate somewhat, and the grass should be dried. This cure works well with small quantities of mediocre grass, since it concentrates the material.

Sun Curing

A quick way to cure small quantities of marijuana is to loosely fill a plastic bag or glass jar, or place a layer between glass or plastic sheets, and expose the material to the sun. Within a few hours the sun begins to bleach it. Turn the marijuana every few hours, so that all parts are exposed to the sun. An even cure is achieved in one to two days (see Plate 16). Some degradation of THC may occur using this method.

Water Cure

Unlike other curing methods, the water cure is performed after the marijuana is dried. Powder and small pieces are most often used, but the cure also works with whole colas. The material is piled loosely in a glass or ceramic pot which is filled with luke-warm water. (When hot water is used, some of the THC is released in oils, which escape and float to the top of the water.) Within a few hours many of the nonpsychoactive water-soluble substances dissolve. An occasional gentle stirring speeds the process. The water is changed and the process repeated. Then the grass is dried again for smoking.

THC is not water-soluble; so it remains on the plant when it is soaked. By eliminating water-soluble substances (pigments, proteins, sugars, and some resins), which may make up 25 percent of the plant material by weight, this cure may increase the concentration of THC by up to a third.

Marijuana cured by this method has a dark, almost black color, and looks twisted and curled, something like tea leaves. The water cure is frequently used to cure dried fan leaves and poor-quality grass.

DRYING

Living marijuana leaves are 80 percent water; colas are about 70 percent water. Marijuana dried for smoking contains only eight to 10 percent water, or about 10 percent of the original amount. There are several methods used to evaporate water; these have little effect on potency, but can affect the taste, bouquet, and smoothness of the smoke. Generally, the slower the dry, the smoother the taste. Excess drying and drying methods that use heat will evaporate some of the volatile oils that give each grass its unique taste and aroma.

Grasses which are dried as part of the curing process usually have a smooth, mild taste, because of the elimination of chlorophyll and various proteins. Cured marijuana may also be a little sweeter than when first picked, because the curing converts some of the plant's starch to simple sugars.

Some grasses are tasty and smooth-smoking when they are dried without curing, especially fresh homegrown buds which retain their volatile oils and sugar. Many homegrowers have acquired a taste for "natural" uncured grass, with its minty chlorophyll flavor; such marijuana is dried directly after harvesting.

Figure 89. Male plants drying on a tree. Cheesecloth holds loose leaf for drying.

Slow Drying

Slow drying is probably the method most commonly used to dry marijuana. Because of the slowness of the dry, a slight cure takes place, eliminating the bite sometimes associated with quickly dried grass.

There are many variations of the technique, but most commonly whole plants or separated colas are suspended upside down from a drawn string or from pegs on a wall in a cool dark room, closet, or other enclosed space. A large number of plants may take a week or two to dry. The drying time for small numbers of plants can be increased (for a slight cure) by placing the plants in large, open paper sacks that have ventilation holes cut in their sides. The drying room should have no heavy drafts, but mold may form on the plants if the air is stagnant. If weather is rainy or the air humid, increase ventilation and watch for mold. Plants should be dried quickly under moderate heat if any mold appears.

Many experienced growers prefer slow drying to curing. There is little chance of error with this method, and buds usually smoke smooth and develop a pliable consistency. Slow-dried ripe buds retain their delicious, sweet aroma and taste.

Figure 90. Slow-drying colas. *Photo by A. Karger.*

Fast Drying

The fast-dry method produces a harsher smoke than slow drying, but it is often the most convenient method to use. The plants are suspended in the same way as for slow drying, but the temperature in the drying area is increased to between 90 and 115 degrees, often by means of electric or gas heaters. The drying area is kept well-ventilated with a fan. As the plants dry, they are removed from the drying area. By this method, plants in a tightly packed room can be dried in less than four days, but the exhaust will contain the deliciously pungent odor of drying marijuana.

Indoor growers often hang plants to dry over radiators or steam pipes. Leaves are dried by placing them on a tray over a radiator or on top of the light fixtures.

Marijuana that is fast-dried retains its original green color and minty taste.

Oven Drying

Oven drying is often used by gardeners to sample their crop. Small quantities of material can be quickly dried by being placed in a 150° to 200° oven for about 10 minutes. Larger quantities can be dried in trays that contain a single layer of material or in a dehydrator. Oven-dried and dehydrator-dried marijuana usually has a harsh taste and bite, and loses much of its bouquet. The method is often used to dry marijuana which has been cured and dried but is too moist to smoke, or to dry marijuana which is to be used for cooking or extractions. It is an adequate method for obtaining dry material for testing and emergencies, but the main harvest should not be dried in this way.

Oven drying works best with leaves. When leaves are dried together with buds or shoots, remove the material from the oven periodically, to separate the faster-drying leaf material (before it burns) from the slower-drying buds. One way to do this is to place all the material on a wire screen over a tray. Every few minutes rub the material across the screen. Dried material falls into the tray and is removed from the oven. Repeat until all the material has dried.

Oven curing works well when closely watched. Dried material that is left in the oven will lose potency quickly. Any time the marijuana begins to char, most of the potency will already have been lost. This should not be a problem unless you are careless, or allow the temperature to go above 200 degrees.

Sun Drying

Some growers dry their crops right in the field. There are many methods of sun drying. In Oregon, some growers break the main stem about two feet from the ground. The leaves and buds dry gradually, since they are still partly attached to the plant. Other growers spread burlap and cover it with plants left to dry. Fan leaves are left on the plants to protect the drying buds from the sun. The grass is manicured after drying. Growers in Arizona shade drying plants with cheesecloth.

Sun-dried marijuana usually has a taste similar to that of oven-dried. Often the sun bleaches it slightly but also destroys some of the delicate bouquet. Prolonged exposure to the sun will decrease potency, although there is no noticeable loss if drying is done quickly.

Dry Ice

Many homegrowers have written to us that the dry-ice cure increases the potency of marijuana considerably, and we would be remiss not to mention it.

Dry ice is frozen carbon dioxide. When it melts (sublimates), it turns from a solid directly into a gas. This gas absorbs some moisture from the frozen marijuana and partially dries it. There are many variations of the dry-ice method. Fresh or partially dried material is usually used, although some enthusiasts claim that the cure also works with dried material. The marijuana is placed in a coffee can or similar container with a lid, along with at least an equal volume of dry ice. Puncture the lid so that the gas can escape as it evaporates. Place the can in a freezer to prolong the evaporation process. When the dry ice is gone, the grass is dried, but still moist.

Some growers claim that simply freezing the grass increases potency. They often freeze fan leaves or other less-potent material for a couple of months before smoking it. This is said to work only with fresh (wet or dried) grass.

FERMENTATION

When vegetation dries, the individual cells which maintained life processes die. But marijuana can still be conditioned by means of fermentation. Fermentation is the process in which microbes and plant enzymes break down complex chemicals into simpler ones, mainly starch and sugars into alcohol and simple acids. In the process chlorophyll is destroyed, giving the material a more ripened appearance. If the fermentation is stopped early, the marijuana has a sweeter taste because of the sugars which the ferment produced.

Fermentation occurs when the moisture content of the marijuana is raised above 15 percent and the temperature is above 60 degrees. The more tightly packed the material, the faster the ferment proceeds. The rate of ferment is controlled primarily by varying the moisture content, but each batch proceeds at its own rate because of differences between plants in nitrogen content. (Nitrogen is necessary to maintain fermenting bacteria.) The process is delicate; should the ferment proceed too rapidly, the marijuana may be converted to compost. Watch the fermentation closely. After the desired color or flavor (from a dried sample) is reached, dry the grass quickly to stop the process.

During fermentation, flavorings can be added to give the

marijuana a spicy aroma. Such spices as cinnamon, cloves, ginger, mace, sage, or vanilla are placed between the fermenting material. Orange, lemon, or lime peels are also used. About half an ounce of spice or four ounces of peel are used for each cubic foot of material to be fermented. The spices are wrapped in cloth sachets. The citrus peels are strung. They can be placed between the layers of marijuana.

There are two types of fermentations: self-generating and forced. They are best used with leaves or immature plants.

Self-Generating Fermentation

Self-generating fermentation proceeds rapidly only when there is enough material to make a heap at least one cubic yard large. When smaller quantities are used, too much of the heat generated by the bacteria is dissipated, so that the process is slow and is more properly considered aging.

Place the material in a large container or in a pile with a tarpaulin placed over it, and lightly spray it with a mister if it is dry. Let the pile heat up for a few days, and then break it down. If it is repacked, the marijuana will develop a dull matte appearance and lose its sugars. If the process is allowed to proceed even further, the marijuana will disintegrate.

Forced Fermentation

Forced fermentation can be used with small quantities of material. It requires an enclosed chamber in which heat and humidity can be regulated.

Pack the marijuana loosely in a kiln or other chamber, and raise the temperature to 135 degrees. Maintain humidity at 75 percent. Check the progress of the ferment periodically. Within a week the ferment should be completed. During this ferment there is a release of ammonia compounds, resulting in some foul odors, but upon completion of the ferment and drying, the marijuana should smoke sweet and mellow.

STORAGE

THC is degraded by both heat and light. Table 26 shows results of an experiment conducted at the University of Mississippi, in

Table 26.

Decomposition of THC in Stored Marijuana[171]

Number of Weeks Elapsed	Percentage of Cannabinoids by Weight in Marijuana[a] Stored at Temperature of									
	0°F		39°F		72°F[b]		97°F		122°F	
	THC	CBN	THC	CBN	THC	CBN	THC	CBN	THC	CBN
Starting	1.3	0.2	1.3	0.2	1.3	0.2	1.3	0.2	1.3	0.2
10									0.75	
20									0.20	
30									0.05	
50	1.25		1.23		1.21					
60							1.03	0.4	0.04	0.73
70							0.29			0.51
100	1.2	c	1.16	c	1.12	c	0.08	0.67		0.41
Percentage of original THC lost	7.69%		10.76%		13.85%		94.00%		97.00%	

[a]Dried, manicured marijuana stored in amber bottles in darkness.
[b]This sample received ambient light.
[c]CBN content increased slightly.

which marijuana was stored under varying temperature conditions.[171] These results indicate that marijuana stored at room temperature (72°) or below and in darkness for up to two years will lose only an insignificant amount of its original potency; whereas marijuana stored in darkness at 97° or above will lose almost all its potency within two years.

In another experiment,[164] Fairbairn stored dried marijuana at different temperatures in both light and dark conditions. The samples in light were exposed to a north-facing window (no direct sunlight). The results are shown in Table 27.

Table 27.

Loss of THC in Marijuana Stored under Various Conditions[164]

	Storage Conditions					
	41° in Darkness		68° in Darkness		68° in Light	
Storage Time	THC Present	Percent of THC Lost	THC Present	Percent of THC Lost	THC Present	Percent of THC Lost
Starting	1.37	0	1.37	0	1.37	0
31 weeks	1.26	8	1.20	12	0.89	35
47 weeks	1.27	7	1.19	12	0.88	36
73 weeks	1.21	11	1.10	20	0.68	50
98 weeks	1.25	9	1.03	25	0.51	63

Fairbairn also performed an experiment to discover the effect of air on THC.[164] Freshly prepared *Cannabis* resin was stored as a loose powder, a compressed powder, and an unbroken lump for one year at 68 degrees F (about room temperature). Samples were stored under two conditions: in light and air, and in darkness and air. The results are shown in Table 28.

Fairbairn experimented further with pure cannabinoids and extracts of marijuana dissolved in petroleum ether, chloroform, and ethanol (alcohol).[165] The results, in Tables 29 and 30, show that THC and CBD in solution are much more unstable than when they are left in marijuana, especially if they are held by the plant in undamaged glands, where they are protected from exposure to air and, to some degree, light. Crude extracts seem more stable than highly refined cannabinoids, especially CBD, which is very unstable in refined solutions.

Table 28.

Loss of THC under Various Conditions of Light and Air[164]

Storage Conditions and Preparation of Sample[a]	THC Present	Percent of THC Lost	CBN
Freshly prepared	11.6	0	Traces
Light and air			
(1) Loose powder	7.3	37	1.8
(2) Compressed broken mass	5.2	55	2.4
(3) Compressed unbroken lump			
(a) Surface layer	5.2	55	2.3
(b) Next 2-mm layer	10.7	8	1.8
(c) Center	11.4	2	1.6
Darkness and air			
(4) Loose powder	12	+3	0.6
(5) Compressed broken mass	10.1	13	1.0
(6) Compressed unbroken lump			
(a) Surface layer	11.1	4	1.9
(b) Next 2-mm layer	10.3	11	1.04
(c) Center	11.0	5	1.47

[a]All samples stored for one year at 68°F.

Extract makers and purchasers should limit the exposure of the solution to light and heat as well as to air. Oils and extracts should be kept refrigerated in opaque, sealed containers. Notice that THC is almost completely degraded in a few weeks when it is held in solution and exposed to light. Red oil, hash oil, and honey oil must be stored in light-tight containers to preserve potency.

From the tables, you can see that light is the primary factor that causes decomposition of THC. The decomposition products are unknown, but are suspected to be polymers or resins. We also do not know whether the rate of decomposition would be faster in direct sunlight.

Air (oxygen) acts much more slowly to convert THC to CBN. Decomposition of THC to CBN is not significant unless temperatures are in the nineties or higher. However, such high temper-

Table 29.

Decomposition of THC in Stored Extracts[164]

Solvent and Storage Time	Percentage of Original THC Present in Sample Stored at		
	41° in Darkness	68° in Darkness	68° in Light
Δ^9-THC in petroleum ether			
Start	100	100	100
2 days	94	95	80
6 days	92	94	9
Δ^8-THC in petroleum ether			
Start	100	100	100
2 days	100	100	51
6 days	99	101	1
THC in $CHCl_3$			
Start		100	100
2 days		99	92
6 days		105	65
20 days		95	9
$CHCl_3$ extract of UNC 255[a]			
Start		100	100
2 days		103	67
6 days		103	15
20 days		101	0
Ethanol extract of UNC 255[a]			
Start		100	100
2 days		100	84
6 days		97	76
20 days		100	55

[a]High THC variety, as tested for THC.

atures can occur in grass that is packed before it is properly dried. The moisture that is left supports microbial activity, which heats the grass internally, as occurs during certain types of curing. Potency of cured grass is not lowered significantly

Table 30.

Decomposition of CBD in Stored Extracts[164]

Solvent and Storage Time	Percentage of Original CBD Present in Sample Stored at	
	68° in Darkness	68° in Light
CBD in $CHCl_3$		
Start	100	100
2 days	69	47
6 days	32	5
20 days	Traces	0
$CHCl_3$ extract of SP1[a]		
Start	100	100
2 days	100	92
6 days	95	61
20 days	103	8
Ethanol extract of SP1[a]		
Start	100	100
2 days	103	80
6 days	106	50
20 days	113	13

[a]High CBD variety, as tested for CBD.

when the cure is done properly and when the buds are left intact during the process.

The figures for powdered and compressed grass in Table 28 show that both light and air cause rapid decomposition when the resin is exposed through breaking of the resin glands. Intact resin glands appear to function well in storing the cannabinoids. For this reason, it is important to handle fresh and dried grass carefully, in order not to crush the material and thus break the glands, especially in the buds, which have a cover of raised resin glands. Most well-prepared marijuana will have intact, well-preserved buds.

The best place to store marijuana is in a dark container in a refrigerator or freezer. *Cannabis* should be stored uncleaned, so

that the glands containing the THC are not damaged, since damage causes their precious contents to be exposed to light and
air. Marijuana should be cleaned only when it is about to be
smoked.

Many growers place a fresh lemon, orange, or lime peel in
with each lid of stored grass. The peel helps to retain moisture,
which keeps the buds pliable, and also gives the grass a pleasant
bouquet.

Most growers take well-earned pride in the quality of the
marijuana that they grow. By supplying yourself with an herb
which may play an important role in your life, you gain a feeling of self-sufficiency that can be infectious.

Since your homegrown is well-tended and fresh, it has a
sweet flavorful taste, far superior to that of commercial grass.
And there need be no fear of contamination from herbicides,
pesticides, adulterants, or other foreign matter. By growing
your own, you come to the pleasant realization that you are
free from the vagaries and paranoia of the marijuana market—
not to mention how little a home garden costs. All of these
feelings can add up to a very heady experience.

In a time of quiet contemplation, you might also reflect on
the experiences that brought you this wondrous herb from a
tiny seed. There is a tradition of mutual nurture and support
between humanity and this plant that goes back 10,000 years.
You are now part of this continuing tradition.

As you probably realized while reading this book, some of
the practical information came to us through letters from growers. We appreciate these letters and will continue to refer to
them when we update and improve future editions of the
Marijuana Grower's Guide. We would also like to hear ideas,
criticisms, and feedback from our readers. Other research material and copies of professional research are also welcome.

Wishing you a Happy Harvest,

Mel Frank
Ed Rosenthal

Bibliographic Notes

HISTORICAL

1. Aldrich, M. A. 1971. A Brief Legal History of Marijuana. Presented to the Western Inst. of Drug Problems Marijuana Conf., Portland, Oregon, Aug. 7, 1971. 14 pp.
2. Frazier, J. 1974. The Marijuana Farmers, Hemp Cults, and Cultures. 133 pp. Solar Age Press. New Orleans, La.
3. Godwin, H. 1967. Pollen Analytic Evidence for the Cultivation of *Cannabis* in England. Rev. Paleobotany Palnyol. 4:71-80.
4. Godwin, H. 1967. The Ancient Cultivation of Hemp. Antiquity 41:42-50, 137-138.
5. Jain, S., and Tarafder, C. 1970. Medicinal Plant Lore of the Santals Tribe in S.E. Asia. Econ. Botany 24:241-249.
6. Kamstra, J. 1974. Weed: Adventures of a Dope Smuggler. 267 pp. Harper and Row. New York.
7. Keng, Hsuan. 1974. Economic Plants of Ancient N. China as Mentioned in *Shih Ching* (Book of Poetry). Econ. Botany 28 (4): 391-410.
8. Li, Hui-Lin. 1974. An Archaeological and Historical Account of *Cannabis* in China. Econ. Botany 28 (4): 437-448.
9. Li, Hui-Lin. 1974. The Origin and Use of *Cannabis* in Eastern Asia: Linguistic-Cultural Implications. Econ. Botany (3): 293-303.
10. Li, Hui-Lin. 1970. The Origin of Plants in S.E. Asia. Econ. Botany 24 (1): 9-10.
11. Musto, D. F. 1972. The Marijuana Tax Act of 1937. Arch. Gen. Psychiat. 26: 101-107.
12. Rea, Mary-Alice F. 1975. Early Introduction of Economic Plants into New England. Econ. Botany 29 (4): 333-356.
13. Rubin, V., and Comitas, L. 1975. Ganja in Jamaica: A Medical Anthropological Study of Chronic Marijuana Use. 206 pp. Moulton/MacFarlane Pub. Scotch Plains, N.J.
14. Schultes, R. E. 1970. Random Thoughts and Queries on the Botany of *Cannabis*. In Joyce, C. R. E., and Curry, S. H., eds. The Botany and Chemistry of *Cannabis*. 241 pp. J. & A. Churchill. London. Pp. 11-38.
15. Schultes, R. E. 1967. Man and Marijuana. Nat. Hist. 82: 59-63, 80, 82.
16. Shakespeare, P. M. 1977. The Book of Pot. 96 pp. A and W Visual Library. New York.

BOTANY

17. Anderson, L. C. 1974. A Study of Systematic Wood Anatomy in *Cannabis*. Botanical Museum Leaflets, Harvard Univ. 24 (2): 29-36.
18. André, Cl., and Vercruysse, A. 1976. Histochemical Study of the Stalked Glandular Hairs of the Female *Cannabis* Plants, Using Fast Blue Salt. Plant Medica 29: 361-366.
19. Bazzaz, F. A., Dusek, D., Sieger, D. S., and Haney, A. W. 1975. Photosynthesis and Cannabinoid Content of Temperate and Tropical Populations of *Cannabis sativa*. Biochem. Syst. Econ. 3: 15-18.
20. Boucher, F., Cosson, L., Unger, J., and Paris, M. 1974. *Cannabis sativa* L.: Chemical Races or Varieties. Plantes Medicinales et Phytother. 8 (1): 20-31. (French)

21. Breslavetz, L. P. 1935. Abnormal Development of Pollen in Different Races and Grafts of Hemp. Genetica 17: 157-170.
22. Clark, R. C. 1977. The Botany and Ecology of Cannabis. 64 pp. Pods Press. Ben Lomond, Calif.
23. Darlington, C. D. 1964. Oil and Fiber Plants. In Chromosomal Botany and the Origin of Cultivated Plants. Allen and Unwin. London. Pp. 182-183.
24. Davidan, G. G. 1972. Hemp. Bull. Applied Bot., Gen., and Plant Breeding 48 (3): 9 pp. Leningrad. (Russian)
25. Dayandan, P., and Kaufman, P. B. 1976. Trichomes of Cannabis sativa (Cannabaceae). Amer. J. Botany 63 (5): 578-591.
26. De Pasquale, A. 1974. Ultrastructure of the Cannabis sativa Glands. Plant Medica 25: 239-248.
27. De Pasquale, A., Tumino, G., and Coasta de Pasquale, R. 1974. Micromorphology of the Epidermic Surfaces of Female Plants of Cannabis sativa L. Bull. Narc. 26 (4): 27-40.
28. Emboden, W. A. 1974. Cannabis, a Polytypic Genus. Econ. Botany 28 (3): 304-310.
29. Emboden, W. A. 1972. Narcotic Plants. 167 pp. Macmillan. New York.
30. Fahn, A. 1967. Laticifers. In Plant Anatomy. Pergamon Press. New York. Pp. 130-134.
31. Fujuta, M., et al. 1967. Studies on Cannabis, II: Examination of the Narcotic and Its Related Components in Hemps, Crude Drugs, and Plant Organs by GLC and TLC. Ann. Rep. Tokyo Coll. Pharm. 17: 238-242. (Japanese)
32. Hammond, C.,T., and Mahlberk, P. G. 1973. Morphology of Glandular Hairs of Cannabis sativa from Scanning Electron Microscopy. Amer. J. Botany 60 (6): 524-528.
33. Haney, A., and Bazzaz, F. A. 1970. Implications of the Distribution of Hemp. In Joyce and Curry (see note 14), pp. 39-48.
34. Heriseset, A., Besson, Ph., and Autin, C. 1965. Toxicological Overview on the Varieties of Hemp Grown in Anjou. Annales Pharm. franc. 23 (11): 631-635. (French)
35. Jenkins, R. W., and Patterson, D. A. 1973. The Relationship Between Chemical Composition and Geographical Origin in Cannabis. Forensic Sci. 2: 59-66.
36. Kechatov, E. A. 1959. Chemical and Biological Evaluation of Resin of Hemp Grown for Seed in Central Districts of the European Part of the USSR. Bull. Narc. 11 (4): 5-9.
37. Lowry, W. T., and Garriott, J. C. 1975. On the Legality of Cannabis: The Responsibility of the Expert Witness. J. Forensic Sci. 20: 624-629.
38. Masoud, A. N., Doorenbos, N. J., and Quimby, W. M. 1973. Mississippi-Grown Cannabis sativa L., IV: Effects of Gibberellic Acid and Indoleacetic Acid. J. Pharm. Sci. 62 (2): 316-318.
39. McPhee, H. C. 1924. Meiotic Cytokinesis of Cannabis. Botan. Gaz. 78: 341-344.
40. Metcalfe, C. R., and Chalk, L. 1969. Cannabinaceae. In Anatomy of the Dicotyledons, II, 1254-1257. Oxford Univ. Press.
41. Miller, N. G. 1970. The Genera of the Cannabaceae in the Southeastern United States. J. Arnold Arboretum 51: 185-195.
42. Nassanov, V. A. 1940. Anatomical Characteristics of Geographical Races of Hemp. Vestnik Solsialisticheskogo Rastenievodstva 4: 102-120. Moscow.
43. Roberts, M. R. 1975. Family Affinities of the Genus Cannabis. Honors Thesis, presented May 1975. Dept. Biology, Harvard Univ. 48 pp.
44. Robinson, B. B. 1944. Marijuana Investigations, IV: A Study of Marijuana Toxicity on Goldfish Applied to Hemp Breeding. J. Amer. Pharm. Assn. 30: 616-619.
45. Schultes, R. E., Klein, W. M., Plowman, T., and Lockwood, T. E. 1974. Cannabis: An Example of Taxonomic Neglect. Botanical Museum Leaflets, Harvard Univ. 23 (9): 337-364.

46. Schultes, R. E. 1970. Random Thoughts and Queries on the Botany of *Cannabis*. *In* Joyce and Curry (see note 14), pp. 11-33.
47. Shoyama, Y., Yagi, M., Nishioka, I., and Yamauchi, T. 1975. Biosynthesis of Cannabinoid Acids. Phytochem. 14 (10): 2189-2192.
48. Simonsen, J. L., and Todd, A. R. 1942. *Cannabis indica*, X: The Essential Oil from Egyptian Hashish. Chem. Society (London) J. (1): 188-191.
49. Small, E. 1975. The Case of the Curious *Cannabis*. Econ. Botany 29 (3): 254.
50. Small, E. 1975. American Law and the Species Problem in *Cannabis:* Science and Semantics. Bull. Narc. 27 (3): 1-17.
51. Small, E., Beckstead, H. D., and Chan, A. 1975. The Evolution of Cannabinoid Phenotypes in *Cannabis*. Econ. Botany 29 (3): 219-232.
    [For similar studies, see the following.
    Lloydia. 36 (2): 144-165.
    Contribution no. 960. 1972. Plant Research Inst., Canadian Dept. Agric. 22 pp.
    Nature. 1973 (Sept.). 245: 147-148. (Summary).]
52. Small, E. 1972. Morphological Variation of Achenes of *Cannabis*. Canad. J. Botany 53 (10): 978-987.
53. Stearn, W. T. 1974. Typification of *Cannabis sativa* L. Botanical Museum Leaflets, Harvard Univ. 23 (9): 325-336.
54. Stearn, W. T. 1970. Botanical Characteristics. *In* Joyce and Curry (see note 14), pp. 1-10.
55. Todd, A. R. 1946. Hashish. Experientia 2 (2): 55-60.
56. Veliky, I. A., and Genest, K. 1972. Growth and Metabolites of *Cannabis sativa* Cell Suspension Cultures. Lloydia 35 (4): 450-456.
57. Warmke, H. E. 1944. Use of Killfish, *Fenulus heteroclitus,* in the Assay of Marijuana. J. Amer. Pharm. Assn. 33: 122-125.
58. Warmke, H. E., and Davidson, H. 1944. Polyploidy Investigations. Carnegie Inst. Wash. Yearbook 43: 153-155.
59. Warmke, H. E., and Davidson, H. 1943-44. Polyploidy Investigations. Carnegie Inst. Wash. Yearbook 43: 135-138.
60. Warmke, H. E. 1941-43. Polyploidy Investigations. Carnegie Inst. Wash. Yearbook 42: 186-187.

CANNABIS CULTIVATION.

61. Agurell, S. 1970. Constituents of Male and Female *Cannabis. In* Joyce and Curry (see note 14), pp. 57-59;
62. Boven, M. V., Bruneel, N., and Daenens, P. 1976. Determination des cannabinoides dans le *Cannabis sativa* d'origine belge. J. Pharm. Belg. 31 (2): 215-219. (French)
63. Coffman, C. B., and Getner, W. A. 1975. Response of *Cannabis sativa* L. to Soil-Applied N, P, and K. (Abstract) Presented to Amer. Soc. of Agron. at Univ. Tenn. Aug. 1975. (S-4): 136.
64. Dewey, L. H. 1913. Hemp. USDA Yearbook, pp. 283-341.
65. Dodge, C. R. 1895. Hemp Culture. USDA Yearbook, pp. 214-223.
66. Doorenbos, N., Fetterman, P., Quimby, M., and Turner, C. E. 1971. Cultivation, Extraction, and Analysis of *Cannabis sativa* L. Ann. N.Y. Acad. Sci., pp. 3-24.
67. Drake, B. 1970. The Cultivators Handbook of Marijuana. 91 pp. Agrarian Reform Co. Eugene, Ore.
68. Fairburn, J. W., and Leibman, J. A. 1974. The Cannabinoid Content of *Cannabis sativa* L. Grown in England. J. Pharmaceut. Pharmacol. 26: 413-419.
69. Fetterman, P., et al. 1971. Mississippi-Grown *Cannabis sativa* L.: Preliminary Observations on Chemical Definition of Phenotype and Variations in THC Content Versus Age, Sex, and Plant Part. J. Pharm. Sci. 60 (8): 1246-1249.
70. Frank, M., and Rosenthal, E. 1974. The Indoor/Outdoor Highest Quality

Marijuana Growers Guide. 94 pp. Level Press, San Francisco, and And/Or Press, Berkeley, Calif.

71. Haney, A., and Kutscheid, B. B. 1973. Quantitative Variation in the Chemical Constituents of Marijuana from Stands of Naturalized *Cannabis sativa* L. in East-Central Illinois. Econ. Botany 27 (2): 193-203.

72. Kimura, M., and Okamoto, K. 1970. Distribution of THC Acid in Fresh Wild *Cannabis*. Specialia 15 (8): 819-20.

73. Krejci, Z. 1970. Changes with Maturation in the Amounts of Biologically Interesting Substances of *Cannabis*. In Joyce and Curry (see note 14), pp. 49-58.

74. Latta, R. P., and Eaton, J. E. 1975. Seasonal Fluctuations in Cannabinoid Content of Kansas Marijuana. Econ. Botany 29: 153-163.

75. Masoud, A. N., and Doorenbos, N. Mississippi-Grown *Cannabis sativa* L., III: Cannabinoid and Cannabinoid Acid Content. J. Pharm. Sci. 62 (2): 313-315.

76. Matchett, J. R., et al. 1940. Marijuana Investigations, II: The Effect of Variety, Maturity, Fertilizer Treatment, and Sex on the Intensity of the Beam Tests. J. Amer. Pharm. Assn 29 (40): 399-404.

77. Nelson, C. H. 1944. Growth Responses of Hemp to Differential Soil and Air Temperatures. Plant Physiol. 19: 295-301.

78. Nordal, A., and Braenden, O. 1973. Variations in Cannabinoid Content of *Cannabis* Plants Grown from the Same Batches of Seeds under Different Ecological Conditions: Preliminary Report. Meddeleser Fra Norsk Farmaceut. Selskap. 35: 8-14.

79. Paris, M., Boucher, F., and Cosson, L. 1975. The Constituents of *Cannabis sativa* Pollen. Econ. Botany 29 (3): 245-253.

80. Phillips, R., et al. 1970. Seasonal Variation in Cannabinolic Acid Content of Indian Marijuana. J. Forensic Sci. 15 (2): 191-200.

81. Quimby, M., Doorenbos, N., Turner, C. E., and Masoud, A. 1975. Mississippi-Grown *Cannabis sativa* L.: Cultivation and Observed Morphological Variations. Econ. Botany 27 (1): 117-127.

82. Richardson, J. 1976. Sinsemilla Marijuana Flowers. 96 pp. And/Or Press. Berkeley, Calif.

83. Robinson, B. B. 1943. Hemp. Farmers' Bull. USDA.

84. Robinson, B. B., and Matchett, J. R. 1940. Marijuana Investigations, III: The Effect of Region of. Growth of Hemp on Response to the Acid and Alkaline Beam Tests. J. Amer. Pharm. Assn 29: 448-453.

85. Schou, J., and Nielson, E. 1970. Cannabinols in Various U.N. Samples and in *Cannabis sativa* Grown in Denmark under Various Conditions. U.N. Sec. ST/SOA/Ser. S/22. 12 pp.

86. Snellen, H. G. 1970. A Quantitative Study of THC during the Growing Season of a Mexican Strain of *Cannabis sativa* L. 40 pp. Honors Thesis no. 13275. Dept. Pharm., Univ. of Miss.

87. Stark, M. 1977. Marijuana Potency (Working Papers). And/Or Press. Berkeley, Calif.

88. Stevens, M. 1975. How to Grow Marijuana Indoors Under Lights. 3d ed. 80 pp. Sun Magic Publishing. Seattle, Wash.

89 Superweed, M. J. 1970. Super Grass Grower Guide. 16 pp. Stone Kingdom Synd. San Rafael, Calif.

90 Superweed, M. J. 1969. The Complete *Cannabis* Cultivator. 16 pp. Stone Kingdom Synd. San Rafael, Calif.

91. Tibeau, Sister Marie Etienne. 1933. Time Factor in Utilization of Mineral Nutrients by Hemp. J. Plant Physiol. 11: 731-747.

92. Turner, C,E., Fetterman, P. S., Hadley, K. W., and Urbanek, J. E. 1975. Constituents of *Cannabis sativa* L., X: Cannabinoid Profile of a Mexican Variant and its Possible Correlation to Pharmacological Activity. Acta Pharm. Jugoslav. 25 (1): 7-16.

SEX DETERMINATION.

93. Arnoux, M., and Mathieu, G. 1969. Influence du milieu sur le phenotype sexuel de descendance Fl issues du croisement entre types dioiques et monoiques de chanvre (*Cannabis sativa* L.). Ann. Amerlior. Plantes 19 (1): 53-58. (French) [For similar studies, see the same journal, 1966, 16 (2): 123-134, and 16 (3): 259-262.]
94. Black, C. A. 1945. Effect of Commercial Fertilizers on the Sex Expression of Hemp. Botan. Gaz. (Sept.), pp. 114-120.
95. Borthwick, H. A., and Scully, N. J. 1954. Photoperiodic Responses of Hemp. Botan. Gaz. (Sept.), pp. 14-29.
96. Helsop-Harrison, J. 1956. Auxin and Sexuality in *Cannabis sativa*. In F. C. Steward, ed. 1971. Plant Physiology, VI, A: 588-597. Academic Press, New York.
97. Laskowska, R. 1961. Influence of the Age of Pollen and Stigmas on Sex Determination in Hemp. Nature 192: 147-148.
98. Mackay, E. L. 1939. Sex Chromosomes of *Cannabis sativa*. Amer. J. Botany 26: 707-708.
99. McPhee, H. C. 1924. The Influence of the Environment on Sex in Hemp, *Cannabis sativa* L. J. Agric. Res. 28: 1067-1080.
100. Molotkovskii, G., and Butnitskii, I. N. 1971. Morphological and Biochemical Characters of Sex in Hemp. Ukr. Botan. Zh. 28 (1): 23-29. (Russian)
101. Menzel, M. Y. 1964. Meiotic Chromosomes of Monoecious Kentucky Hemp (*Cannabis sativa*). Bull. Torrey Botan. Club 91 (3): 193-205.
102. Pritchard, F. J. 1916(7). Change of Sex in Hemp. J. Heredity 7: 325-329.
103. Ram, H. Y., and Jaiswal, V. S. 1972. Induction of Male Flowers on Female Plants of *Cannabis sativa* by Gibberellins and its Inhibition by Abscisic Acid. Planta 105: 263-266.
104. Ram, H. Y., and Jaiswal, V. S. 1972. Sex Reversal in the Male Plants of *Cannabis sativa* L. by Ethyl Hydrogen-1-Propylphosphate. Z. Pflanzenphysiol. 68: 181-182.
105. Schaffner, J. H. 1931. The Fluctuation Curve of Sex Reversal in Staminate Hemp Plants Induced by Photoperiodicity. Amer. J. Botany 18: 324-430. [For similar studies, see Ohio J. Sci. 25 (1925): 172-176; Ecology 4 (1923): 323-324; Botan. Gaz. 71 (1921): 197-291.]
106. Schaffner, J. H. 1926. The Change of Opposite to Alternate Phylotaxy and Repeated Rejuvenations in Hemp by Means of Changed Photoperiodicity. Ecology 7 (3): 315-325.
107. Small, E. 1972. Interfertility and Chromosomal Uniformity in *Cannabis*. Canad. J. Botany 50 (2): 1947-1948.

CHEMISTRY.

108. Allwardt, W. H., Babcock, P. A., Segelman, A. B., and Cross, J. M. 1972. Photochemical Studies of Marijuana (*Cannabis*) Constituents. J. Pharm. Sci. 61 (12): 1994-1996.
109. Augrill, S., and Nillson, J. L. G. 1972. The Chemistry and Biological Activity of *Cannabis*. Bull. Narc. 24 (4): 2935-2937.
110. Crombie, L., and Crombie, W. M. L. 1975. Cannabinoid Bis-Homologues: Miniaturized Synthesis and GLC Study. Phytochem. 14: 213-220.
111. Davis, H. K., Jr., et al. 1970. The Preparation and Analysis of Enriched and Pure Cannabinoids from Marijuana and Hashish. Lloydia 33 (4): 453-460.
112. Duke, E. L., and Reimann, B. E. F. 1973. The Extractability of the Duquenois-Positive Cannabinoids. Toxicol. 1: 289-300.
113. Edery, H., Grunfield, Y., Ben-Zvi, Z., and Mechoulam, R. 1971. Structural Requirements for Cannabinoid Activity. Ann. N.Y. Acad. Sci.: 40-53.

114. Fairburn, J. W., and Leibman, J. A. 1973. The Extraction and Estimation of Cannabinoids in *Cannabis sativa* L. and its Products. J. Pharmaceut. Pharmacol. 25: 150-153.
115. Fiechtl, J., and Spiteller, G. 1975. New Cannabinoids: Part I. Tetrahedron 31 (6): 479-488. (German with English summaries)
116. Grinspoon, L. 1969. Marijuana. *In* Altered States of Awareness: Readings from Scientific American, pp. 89-98. W. H. Freeman. San Francisco, Calif.
117. Grlic, Lj. 1964. A Study of Some Chemical Characteristics of the Resin from Experimentally Grown *Cannabis* from Different Origins. United Nations Sec. ST/SOA/Ser. S/10. 6 pp.
118. Harvey, D. J. 1976. Characterization of the Butyl Homologues of THC, CBN, and CBD in samples of *Cannabis* by Combined Gas Chromatography and Mass Spectrometry. J. Pharmaceut. Pharmacol. 28 (4): 280-285.
119. Holley, J. H., Hadley, K. W., and Turner, C. E. 1975. Constituents of *Cannabis sativa* L., XI: Cannabidiol and Cannabichromene in Samples of Known Geographical Origin. J. Pharm. Sci. 64 (5): 892-894.
120. Masoud, A. N., and Doorenbos, N. J. 1973. Mississippi-Grown *Cannabis sativa* L., III: Cannabinoid and Cannabinoid Acid Content. J. Pharm. Sci. 62 (2): 313-315.
121. Mechoulam, R., McCallum, N. K., and Burstein, S. 1976. Recent Advances in the Chemistry and Biochemistry of *Cannabis*. Chem. Rev. 76 (1): 75-110.
122. Mechoulam, R. 1973. Marijuana Chemistry, Pharmacology, Metabolism, and Clinical Effects. 409 pp. Academic Press. New York.
123. Mechoulam, R. 1970. Marijuana Chemistry. Science 168 (3936): 1159-1166.
124. Mechoulam, R., and Gaoni, Y. 1967. Recent Advances in the Chemistry of Hashish. Fortschritte Chem. Organ. Naturwiss. 25: 175-213.
125. Mechoulam, R., and Gaoni, Y. 1965. Hashish, IV: The Isolation and Structure of Cannabinolic, Cannabidiolic, and Cannabigerolic Acids. Tetrahedron 121: 1223-1229.
126. Merkis, F. 1971. Cannabivarin and Tetracannabivarin: Two New Constituents of Hashish. Nature 232: 579-580.
127. Mole, Jr., L., Turner, C. E., and Henry, J. T. 1974. Delta-9-Tetrahydrocannabinolic Acid "B" from an Indian Variant of *Cannabis sativa*. U.N. Sec. ST/SOA/Ser. S/48. 8 pp.
128. Moreton, J. E., and Davis, W. M. 1972. A Simple Method for Preparation of Injectables of THC and *Cannabis* Extracts. J. Pharmaceut. Pharmacol. 61 (1): 176-177.
129. Poddar, M. K., Ghosh, J. J., and Datta, J. 1975. A Micromethod for the Estimation of *Cannabis* Components. Science and Culture 41 (10): 492-494.
130. Radosevic, A., Kupinic, M., and Grlic, L. 1962. Antibiotic Activity of Various Types of *Cannabis* Resin. U.N. Sec. ST/SOA/Ser. S/6. 12 pp.
131. Shani, A., and Mechoulam, R. 1971. Photochemical Reactions of CBD: Cyclization to THC and Other Transformations. Tetrahedron 27 (2): 601-606.
132. Straight, R., Wayne, A. W., Lewis, E. G., and Beck, E. C. 1973. Marijuana Extraction and Purification for Oral Administration of Known Amounts of Delta-9-Tetrahydrocannabinol (THC). Biochem. Medicine 8: 341-344.
133. Toffoli, F., Avico, U., and Ciranni, E. 1968. Methods of Distinguishing Biologically Active *Cannabis* and Fiber Materials. Bull. Narc. 20 (1): 55-59.
134. Turner, C. E., et al. 1975. Constituents of *cannabis sativa* L., VIII: Possible Biological Application of a New Method to Separate CBD and CBC. J. Pharm. Sci. 64 (5): 810-814.
135. Turner, C,E., and Hadley, K. W. 1974. Chemical Analysis of *Cannabis sativa* of Distinct Origin. Arch. de Invest. Med. 5 (1): 135-140.
136. Turner, C. E., Hadley, K. W., and Fetterman, P. 1973. Constituents of *Cannabis sativa* L., VI: Propyl Homologues in Samples of Known Geographical Origin. J. Pharm. Sci. 62 (10): 1739-1741.

137. Turner, C. E., and Hadley, K. W. 1973. Constituents of *Cannabis sativa* L., III: Clear and Discreet Separation of CBD and CBC. J. Pharm. Sci. 62 (7): 1083-1086.

138. Turner, C. E. and Hadley, K. W. 1973. Constituents of *Cannabis sativa* L., II: Absence of CBD in an African Variant. J. Pharm. Sci. 62 (2): 251-255.

139. Vree, T. B., Breimer, D. D., Van Ginneken, C. A. M., and Van Rossum, J. M. 1972. Identification in Hashish of THC, CBD, and CBN Analogues with a Methyl Sidechain. J. Pharmaceut. Pharmacol. 24: 7-12.

140. Wood, A., *et al.* 1896. Charas: The Resin of Indian Hemp. Chem. Soc. (London) J., pp. 539-546.

NONCANNABINOIDS OF *CANNABIS*.

141. El-Feraly, F. S., and Turner, C. E. 1975. Alkaloids of *Cannabis sativa* Leaves. Phytochem. 14 (10): 2304.

142. Ham, M. T., *et al.* 1973. Effects of *Cannabis* Roots on the Heart. J. Amer. Med. Assn (Letters) 225 (5): 525.

143. Hendricks, H., Malingre, T. M., Batterman, S., and Bos, R. 1975. Mono- and Sesqui-Terpene Hydrocarbons of the Essential Oil of *Cannabis sativa*. Phytochem. 14 (3): 814-815.

144. Hood, L. V. S., Dames, M. E., and Barry, G. T. 1973. Headspace Volatiles of Marijuana. Nature 242: 402-404.

145. Lotter, H. L., and Abraham, D. J. 1975. Cannabisativine: A New Alkaloid from *Cannabis sativa* L. Root. Tetrahedron Letters 33: 2815-2818.

146. Malingre, T., *et al.* 1975. The Essential Oil of *Cannabis sativa*. Planta Medica 28: 56-61.

147. Martin, L., Smith, D. M., and Farmilo, C. G. 1961. Essential Oil from *Cannabis sativa* and its Use in Identification. Nature 191: 774-776.

148. Mobarak, Z., Bieniek, D., and Korte, F. 1974. Studies on Non-Cannabinoids of Hashish, II. Chemosphere 6: 265-270.

149. Mole, M. L., Buelke, J., and Turner, C. E. 1974. Preliminary Observations on Cardiac Activities of *Cannabis sativa* Root Extracts. J. Pharm. Sci. 63 (7): 169-170.

150. Mole, M. L., and Turner, C. E. 1974. Phytochemical Screening of *Cannabis sativa* L., I: Constituents of an Indian Variant. J. Pharm. Sci. 63 (1): 154-156.

151. Mole, M. L., and Turner, C. E. 1973. Phytochemical Screening of *Cannabis sativa* L., II: Choline and Neurine in the Roots of a Mexican Variant. Acta Pharm. Jugoslav. 23 (4): 203-205.

152. Nigam, M. C., Handa, K. L., Nigam, I. C., and Levi, L. 1965. Essential Oils of Marijuana: Composition of Genuine Indian *Cannabis sativa* L. Canad. J. Chem. 43 (4): 3372-3376.

153. Rogers, J. R. 1971. *Cannabis* Roots. J. Amer. Med. Assn (Letters) 217 (12): 1705-1706.

154. Rybicka, H., and Engelbrecht, L. 1974. Zeatin in *Cannabis* Fruit. Phytochem. 13: 282-283.

155. Slatkin, D. J., *et al.* 1975. Steroids of *Cannabis Sativa* Root. Phytochem. 14: 58-581.

156. Slatkin, D. J., *et al.* 1971. Chemical Constituents of *Cannabis sativa* L. Root. J. Pharm. Sci. 60 (12): 1891-1892.

157. St. Angelo, A., and Ory, R. L. 1970. Properties of a Purified Proteinase from Hempseed. Phytochem. 9: 1933-1938.

158. Stahl, E., and Kunde, R. 1973. Neue Inhaltsstoffe aus dem Ätherischen Öl von Cannabis sativa. *Tetrahedron Letters 30: 2841-2844. (German)*

159. Turner, C. E., and Mole, M. L. 1973. Chemical Components of *Cannabis sativa*. J. Amer. Med. Assn (Letters) 225 (6): 639.

160. Wold, J. K., and Hillstad, A. 1976. Demonstration of Galactosoamine in a Higher Plant: *Cannabis sativa.* Phytochem. 15 (2): 325-326.

STORAGE/SMOKING.

161. Agurell, S., and Leander, K. 1971. Stability, Transfer, and Absorption of Cannabinoid Constituents of *Cannabis* (Hashish) during Smoking. Acta Pharm. Suevica 8: 391-402.
162. Chiesa, E. P., Rondina, R. V. D., and Couisso, J. D. 1973. Chemical Composition and Potential Activity of Argentine Marijuana. J. Pharmaceut. Pharmacol. 25: 953-956.
163. Coffman, C. B., and Gentner, W. A. 1974. *Cannabis sativa* L.: Effect of Drying Time and Temperature on Cannabinoid Profile of Stored Leaf Tissue. Bull. Narc. 26 (1): 68-70.
164. Fairburn, J. W., Liebman, J. A., and Rowan, M. G. 1976. The Stability of *Cannabis* and its Preparations on Storage. J. Pharmaceut. Pharmacol. 28 (1): 1-7.
165. Jones, L. A., and Foote, R. S. 1975. *Cannabis* Smoke Condensate: Identification of Some Acids, Bases, and Phenols. J. Agric. Food Chem. 23 (6): 1129-1131.
166. Kinzer, W., *et al.* 1974. The Fate of the Cannabinoid Components of Marijuana during Smoking. Bull. Narc. 26 (3): 41-53.
167. Koehler, B. 1946. Hemp-Seed Treatment in Relation to Different Dosages and Conditions of Storage. Phytopathol. 36: 937-943.
168. Patel, A. R., and Gori, G. B. 1975. Preparation and Monitoring of Marijuana Smoke-Condensate Samples. Bull. Narc. 27 (3): 47·54.
169. Savaki, H. E., Cunah, J. Carlini, E. A., and Kepholas, T. A. 1976. Pharmacological Activity of Three Fractions Obtained by Smoking *Cannabis* through a Water Pipe. Bull. Narc. 28 (2): 49-56.
170. Turner, C. E., and Hadley, K. W. 1975. Constituents of *Cannabis sativa* L., IX: Stability of Synthetic and Naturally Occurring Cannabinoids in Chloroform. J. Pharm. Sci. 64 (2): 357-359.
171. Turner, C. E., *et al.* 1973. Constituents of *Cannabis sativa* L., IV: Stability of Cannabinoids in Stored Plant Material. J. Pharm. Sci. 62 (10): 1601-1605.
172. Turner, C. E., Hadley, K. W., and Davis, K. H. 1973. Constituents of *Cannabis sativa* L., V: Stability of an Analytical Sample Extracted with Chloroform. Acta Pharm. Jugoslav. 23 (2): 89-94.
173. Turner, C. E., and Hadley, K. W. 1973. Preservation of *Cannabis.* J. Amer. Med. Assn (Letters) 223 (9): 1043.

PHARMACOLOGY.

174. Agurell, S., *et al.* 1973. Quantitation of THC in Plasma from *Cannabis* Smokers. J. Pharmaceut. Pharmacol. 25: 554-558.
175. Agurell, S., and Nillson, J. L. G. 1972. The Chemistry and Biological Activity of *Cannabis.* Bull. Narc. 24 (4): 2935-2937.
176. Agurell, S. 1970. Chemical and Pharmacological Studies of *Cannabis. In* Joyce and Curry (see note 14), pp. 175-191.
177. Christensen, H. D., *et al.* 1975. Activity of Delta-8- and Delta-9-THC and Related ComPounds in the Mouse. Science 31: 165-167.
178. Davis, W. M. J. 1972. A Simple Method for the Preparation of Injectables of THC and *Cannabis* Extracts. J. Pharmaceut. Pharmacol. 24: 176.
179. Edery, H., Grunfeld, Y., Ben-Zvi, Z., and Mechoulam, R. 1971. Structural Requirements for Cannabinoid Activity. Annals N.Y. Acad. Sci: 40-53.
180. Gill, E. W., and Paton, W. D. M. 1970. Pharmacological Experiments *in vitro* on

the Active Principles of *Cannabis*. *In* Joyce and Curry (see note 14), pp. 165-173.

181. Gill, E. W., Paton, W. D. M., and Pertwee, R. G. 1970. Preliminary Experiments on Chemistry and Pharmacology of *Cannabis*. Nature 228: 134-136.

182. Haagen-Smit, A. J., *et al.* A Physiologically Active Principle from *Cannabis sativa* (Marijuana). Science 91 (2372): 602-603.

183. Hollister, L. E., and Gillespie, H. 1975. Interactions in Man of THC, II: CBN and CBD. Clin. Pharm. Therap. 18: 80-83.

184. Kabelik, J., Krejci, Z., and Santavy, F. 1960. *Cannabis* as a Medicant. Bull. Narc. 12 (3): 5-23.

185. Karniol, I. G., *et al.* 1975. Effects of THC and CBN in Man. Pharmacol. 13: 502-512.

186. Karniol, I. G., and Carlini, E. A. 1973. Pharmacological Interaction between CBD and THC. Psychopharmacologia 33: 53-70.

187. Lemberger, L., *et al.* 1971. THC: Metabolism and Disposition in Long-Term Marijuana Smokers. Science 173: 72-74.

188. Maugh, T. H. 1974. Marijuana: The Grass May No Longer Be Greener. Science 185: 683-685.

189. McCallum, N. K. 1975. The Effect of CBN on Δ^1 THC Clearance from the Blood. Experientia 31 (8): 957-958.

190. Olsen, J. L., *et al.* 1975. An Inhalation Aerosol of THC. J. Pharmaceut. Pharmacol. (Communications) 28: 86-97.

191. Paton, W. D. M. Pharmacology of Marijuana. 1975. Ann. Rev. Pharm. 15: 191-219.

192. Perez-Reyes, M. Timmons, M. C., Davis, K. H., and Wall, E. M. 1973. A Comparison of the Pharmacological Activity in Man of Intravenously Administered Delta-nine-THC, CBN, and CBD. Experientia 29: 1368-1369.

193. Powel, G., Salmon, M., and Bembry, T. H. 1941. The Active Principle of Marijuana. Science 93 (2422): 521-522.

194. Mikuriya, T. H. 1973. Marijuana: Medical Papers. 475 pp. Medi-Comp Press, Oakland, Calif.

195. Sallen, S. E., *et al.* 1975. Antiemetic Effect of THC in Patients Receiving Cancer Chemotherapy. New England Med. J. 293 (16): 795-798.

196. Segelman, A. B., and Sofia, R. D. 1973. *Cannabis sativa* L. (Marijuana), IV: Chemical Basis for Increased Potency Related to a Novel Method of Preparation. J. Pharm. Sci. 62 (12): 2044-2046.

197. Turner, C. E. 1974. Active Substances in Marijuana. Arch. de Invest. Medica 5 (1): 135-140.

MISCELLANEOUS.

198. Bear, F. E. 1942. Soils and Fertilizers. 374 pp. John Wiley and Sons. New York.

199. Bickford, E. D., and Dunn, S. 1972. Lighting for Plant Growth. 221 pp. Kent State Univ. Press. Kent, Ohio.

200. Black, C. A. 1968. Soil Plant Relationships. 2d ed. John Wiley and Sons. New York.

201. Bradshaw, A. D. 1972. Some of the Evolutionary Consequences of Being a Plant. Evolutionary Biology 5: 25-47. Appleton-Century-Crofts.

202. Canham, A. E. 1966. Artificial Light in Horticulture. 212 pp. Centrex Pub. Einhoven, the Netherlands.

203. Chichilo, P., and Whittaker, C. W. 1961. Trace Elements in Agricultural Limestones of the United States. Agronomy J. 53 (3): 143-44.

204. Common Weeds of the United States. 1970. 463 pp. USDA. Washington, D.C.

205. Crombie, L., and Crombie, W. M. L. 1975. Cannabinoid Formation in *Cannabis sativa* Grafted Inter-Racially, and with Two *Humulus* Species. Phytochem. 14: 409-412.

206. Devlin, R. M. 1969. Plant Physiology. 2d ed. 446 pp. Van Nostrand Reinhold. New York.
207. DeWit, J. M. J., and Harlan, J. R. 1975. Weeds and Domesticates: Evolution in the Man-Made Habitat. Econ. Botany 29 (2): 99-107.
208. Donahue, R. L., Shickluna, J. C., and Robertson, L. S. 1971. Soils: An Introduction to Soils and Plant Growth. 3d ed. 587 pp. Prentice-Hall. Englewood Cliffs, N.J.
209. Duke, W. B., Hagin, R. D., Hunt, J. F., and Lindscott, D. L. 1975. Metal Halide Lamps for Supplemental Lighting in Greenhouses: Crop Responses and Spectral Distribution. Agronomy J. 67: 49-53.
210. Ehrlich, P. R., and Rayen, P. H. 1969. Differentiation of Populations. Science 165: 1228-1231.
211. Encyclopaedia Brittanica. 1894. Vol. III: 627-628. Vol. XI: 647-649, 843. Vol. XVII: 231-233, 739-745.
212. Fenselau, C., Kelly, S., Salmon,· M., and Billets, S. 1976. The Absence of THC from Hops. Food Socmet. Toxicol. 14 (1): 35-39.
213. Getting the Bugs Out of Organic Gardening. 1973. 115 pp. Rodale Press. Emmaus, Pa.
214. Golueke, C. G. 1972. Composting: A Study of the Process and its Principles. 110 pp. Rodale Press. Emmaus, Pa.
215. Hall, A. D. 1912. Fertilizers and Manures. 384 pp. E. P. Dutton. New York.
216. Harlan, J. R. 1976. The Plants and Animals that Nourish Man. Scientific Amer. (Sept.): 89-97.
217. Henis, C. Y., and Mitchell, R. 1973. Effect of Biogenic Amines and Cannabinoids on Bacterial Chemotaxis. Bacteriol. 115 (3): 1215-1218.
218. Lentz, P. L., Turner, C. E., Robertson, L. W., and Getner, W. A. 1974. First American Record of Cercospora Cannabina, with Notes on the Identification of C. cannabina and C. cannabis. Plant Disease Reporter 58 (2): 165-168.
219. Levin, D. A. 1973. The Role of Trichomes in Plant Defense. Q. Rev. Biol. 48 (1): 3-15.
220. McVickar, M. H., Bridger, G. L., and Nelson, L. B., eds. 1963. Fertilizer Technology and Usage. 464 pp. Soil Society of America. Madison, Wisc.
221. Organic Fertilizers: Which Ones and How to Use Them. 1973. 129 pp. Rodale Press. Emmaus, Pa.
222. Rorison, I. H., ed. 1969. Ecological Aspects of Mineral Nutrition in Plants. 484 pp. Blackwell Sci. Pub., no. 81 Oxford, England.
223. Stevenson and Brown. 1918. Soil Survey of Iowa Report No. 2, Pottawattamie County Soils. 27 pp. Iowa State College of Agriculture and Mechanical Arts. Ames, Iowa.
224. Stout, R., and Clemence, R. 1971. The Ruth Stout No-Work Garden Book. 218 pp. Rodale Press. Emmaus, Pa.
225. Sunset Western Garden Book. 1967. 448 pp. Lane Publishing. Menlo Park, Calif.
226. Teuscher, H., and Adler, D. 1960. The Soil and its Fertility. 446 pp. Reinhold Publishing. New York.
227. Thompson, L. 1957. Soils and Fertility. 2d ed. 327 pp. McGraw-Hill. New York.
228. Vegetable Gardening. 16th ed. 1972. 72 pp. Lane Books. Menlo Park, Calif.
229. Wheeler, H. J. 1924. Manures and Fertilizers. 389 pp. Macmillan. New York.
230. Yepsen, R. B., Jr., ed. 1976. Organic Plant Protection. 688 pp. Rodale Press. Emmaus, Pa.
231. Coffman, C. B., and Gentner, W. A. 1975. Cannabinoid Profile and Elemental Uptake of Cannabis sativa L. as Influenced by Soil Characteristics. Agronomy J. 67: 491-497.
232. Shearer, T. M., and Chalmers, D. 1970. Tobacco Growing in the British Isles. 31 pp. Scottish Amateur Tobacco Growing and Curing Assn. Kirkcoldy, Scotland.

Index

The *FITZ HUGH LUDLOW MEMORIAL LIBRARY* in San Francisco houses one of the world's finest collections of rare drug books, manuscripts, and ephemera. The archive totals more than 11,000 items ranging from inscribed first editions to obscure mimeo handouts, with emphasis on literary and historical works on cannabis, opiates, cocaine, hallucinogens and anesthetics. Special collections include materials on drug magic, mysticism and folklore; personal memoirs and biographies; writings of the beat and hippie generations; drugs and music; drugs and sex; and drug art, photography, and paraphernalia. The Library is open to members as a non-circulating reference collection and is intended to serve as a resource center for advanced research in psychoactive drug history, literature and lore.

Memberships and donations help the Library pay the rent and expand the collection. All members receive the Library newsletter which appears sporadically, and the staff will try to help members with specific research questions as time permits. Memberships start at $10 a year for "Corresponding Members" who want occasional help in locating references by mail, and $25 a year for "Research Members" who need to use the collection more extensively. In lieu of membership fees, the Library will happily accept donations of inscribed books and offprints, manuscripts, artworks, mounted photographs, unusual artifacts, phonograph records, and other items (no drugs please!) not already found in the archives. For further information, please write to The Fitz Hugh Ludlow Memorial Library, P.O. Box 99346, San Francisco, California 94109.

SINSEMILLA: Marijuana Flowers
by Jim Richardson
Photography by Arik Woods
Foreword by David Crosby

**$9.95, 96 pages, 8½x11, perfectbound
over 120 full color photographs
ISBN: 0-915904-23-3**

"If you enjoy gazing upon dream dope in all its flowering glory, this book is full of pinups any centerfold would be proud to display. It's a shame that seeing isn't smoking."

—High Times Magazine

"A subject of never-ending interest is beautifully photographed in the book entitled *SINSEMILLA: MARIJUANA FLOWERS.*"

—The SF Review of Books

SINSEMILLA, a pictorial journey of the genesis of the marijuana plant from the germination of the seed to the ripening of the flowers, has been the subject of widespread critical acclaim. The revealing text by the farmers is coupled with over 120 full color illustrations, forming a lyric marijuana odyssey.

There are now 40,000 copies of *Sinsemilla* in print.

MARIJUANA POTENCY
by Michael Starks

$4.95, 128 pages, 5½x8½, perfectbound illustrated
ISBN: 0-915904-27-6

MARIJUANA POTENCY is the best current book dealing with the active constituents of marijuana, their effects, and the latest techniques for testing and increasing potency.

Includes an account of classical and modern techniques for making hasheesh, hash oil and other potent preparations.

Michael Starks surveys the academic literature as well as black market studies of isomerization of inactive into active compounds, and the chemical synthesis of THC and related compounds. Exceptional illustrations augment the usefulness of this work.

MARIJUANA POTENCY is an essential reference for researchers and marijuana aficionados.

CAMP is scheduling Pot Smokers Rights Demonstrations, Reefer Rallies and Marijuana Marches around the country. The strategy is based on the observation that 5000 protestors are more difficult for a legislator to ignore than 5000 letters.

CAMP welcomes all pro-pot organizations and individuals to joint with us in the campaign to abolish pot prohibition.

The Marijuana Movement has grown from a few seeds of discontent into a vigorous national campaign to totally abolish pot prohibition. Heads are weary of writing legislators and are taking to the streets. (The "amotivational syndrome" touted for so long be government-funded researchers goes up in smoke.)

Coalition for the Abolition
of Marijuana Prohibition
P.O. Box 53265
Atlanta, Georgia 30355

NAME_____

Street_____

City/State_____ Zip Code_____

I'm joining CAMP and want to pay my dues. Here's my $10 annual fee. Send along my CAMPer's Guide, CAMP Button, and more information about CAMP.
Enclosed is a $_____ donation for the CAMP Legal Defense Fund.
I need more CAMP gear: _____ CAMP Buttons _____ CAMP Tee-Shirts
@ 50¢ Each @ $6 Each
(State Size)
Additional information: Please enclose 50¢ postage and handling. Please allow 4–6 weeks for delivery.

THE AMERICAN HARVEST COMMITTEE

The American Harvest Committee is a coalition formed to work toward the elimination of all criminal penalties for victimless actions. The organization has promoted the Right-To-Harvest Festival in the San Francisco Bay Area and is currently expanding its activities to support other groups throughout the nation. The organization is entirely self-supporting and relies on the sales of its tee-shirts for funds to continue the fight for freedom.

Right-To-Harvest Tee-Shirt
4 Color All-Cotton Silk-Screen
$6.50 Including Postage & Handling

Send check or Money Order to
American Harvest Committee
Box 531
625 Post Street
San Francisco, Ca. 94102

FREEDOM IS THE ISSUE